DIFFERENTIAL EQUATIONS
AND APPLICATIONS

NORTH-HOLLAND
**MATHEMATICS STUDIES** **31**

---

# Differential Equations and Applications

Proceedings of the Third Scheveningen Conference on Differential Equations,
The Netherlands, August 29 - September 2, 1977

Edited by
**WIKTOR ECKHAUS**
*University of Utrecht*
and
**EDUARD M. DE JAGER**
*University of Amsterdam*

1978

NORTH-HOLLAND PUBLISHING COMPANY
AMSTERDAM · NEW YORK · OXFORD

*ISBN 0 444 85185 2*

*Publishers*

NORTH-HOLLAND PUBLISHING COMPANY
AMSTERDAM • NEW YORK • OXFORD

*Sole distributors for the U.S.A. and Canada*
ELSEVIER NORTH-HOLLAND, INC.
52 VANDERBILT AVENUE,
NEW YORK, N.Y. 10017

PRINTED IN THE NETHERLANDS

# PREFACE

This volume is an account of the lectures delivered at the third Scheveningen Conference on Differential Equations.

The conference was again organized by B.L.J. Braaksma (University of Groningen), W. Eckhaus (University of Utrecht), E.M. de Jager (University of Amsterdam) and H. Lemei (Technical University of Delft), and was made possible through the financial support of the Minister of Education and Sciences of the Netherlands.

Like the two preceding conferences (North-Holland Mathematics Studies, Vols. 13 and 21) the aim of this conference was to gather a number of mathematicians actively engaged in research in the field of differential equations and thus obtain an up to date picture of the recent developments.

The lectures presented in this volume cover a large variety of aspects of differential equations, and range from abstract theory to applications. Following topics may be mentioned a.o.: non-linear differential equations of elliptic, parabolic and hyperbolic type, upper and lower bounds of solutions, non-linear integral equations, free boundary problems, singular perturbations and control theory. Further, applications in fluid mechanics, diffusion problems, wave mechanics, transport theory and epidemiology.

It is a pleasure to acknowledge the gratitude to all authors for their contributions.

Wiktor Eckhaus
Eduard M. de Jager, Editors
Utrecht, Amsterdam, March, 1978.

CONTENTS

## LIST OF PARTICIPANTS

Invited speakers

| | |
|---|---|
| H. Amann | Ruhr Universität, Bochum, Germany |
| D.L. Colton | University of Strathclyde, Glasgow, Scotland |
| G. Duvaut | Université de Paris, VI , France |
| W.N. Everitt | University of Dundee, Scotland |
| G. Fichera | University of Rome, Italy |
| K. Kirchgässner | Universität Stuttgart, Germany |
| J.A. Nohel | University of Wisconsin, Madison, (U.S.A.) |
| P. Rabinowitz | University of Wisconsin, Madison, (U.S.A.) |
| J. Schröder | Universität zu Köln, Cologne, Germany |
| R. Temam | Université de Paris, Orsay, France |

Other participants

| | |
|---|---|
| B.L.J. Braaksma | Rijksuniversiteit Groningen |
| A.H. v.d. Burgh | Rijksuniversiteit Utrecht |
| Ph. Clement | Technische Hogeschool Delft |
| T.M.T. Coolen | Universiteit van Amsterdam |
| C. Cuvelier | Technische Hogeschool Delft |
| O. Diekman | Mathematisch Centrum, Amsterdam |
| C.J. van Duyn | Rijksuniversiteit Leiden |
| A. Dijksma | Rijksuniversiteit Groningen |
| W. Eckhaus | Rijksuniversiteit Utrecht |
| B.A. Fleishman | Renselaer Polytechnic, Troy, U.S.A. |
| J.A. van Gelderen | Technische Hogeschool, Delft |
| J. de Graaf | Technische Hogeschool Delft |
| J. Grasman | Mathematisch Centrum, Amsterdam |
| P.P.N. de Groen | Technische Hogeschool Eindhoven |
| Br. van Groesen | Technische Hogeschool Eindhoven |
| R.J. Hangelbroek | Kath. Universiteit Nijmegen |
| A. van Harten | Rijksuniversiteit Utrecht |
| M.H. Hendriks | Landbouwhogeschool Wageningen |
| D. Hilhorst | Mathematisch Centrum, Amsterdam |
| E.M. de Jager | Universiteit van Amsterdam |
| M.J.W. Jansen | Vrije Universiteit Amsterdam |
| J.J.I.M. van Kan | Technische Hogeschool Delft |
| E.W.M. Koper | Universiteit van Amsterdam |
| R. Martini | Technische Hogeschool Delft |
| J.J.H. Miller | Kath. Universiteit Nijmegen |
| H. Moet | Rijksuniversiteit Utrecht |
| G.Y. Nieuwland | Vrije Universiteit Amsterdam |
| L.A. Peletier | Rijksuniversiteit Leiden |
| Mevr. J. Potters | Vrije Universiteit Amsterdam |
| A.M. Reuvers | Universiteit van Amsterdam |
| J.W. Reijn | Techn. Hogeschool Delft |
| J.W. de Roever | Mathematisch Centrum Amsterdam |
| J.M. Schumacher | Vrije Universiteit Amsterdam |
| J. Sijbrand | Rijksuniversiteit Utrecht |
| E.J.M. Veling | Mathematisch Centrum Amsterdam |
| H. Lemei | Technische Hogeschool Delft. |

*Differential Equations and Applications*
*W. Eckhaus and E.M. de Jager (eds.)*
*©North-Holland Publishing Company (1978)*

# EXISTENCE AND STABILITY FOR SOME SYSTEMS OF DIFFUSION-REACTION EQUATIONS

Herbert Amann
Mathematisches Institut der Ruhr-Universität
Bochum, Germany

In this paper we consider initial boundary value problems for semilinear systems of parabolic equations. This research has been motivated by the following system:

$$
(1)\qquad
\begin{aligned}
&\frac{\partial u}{\partial t} - \Delta u = -\alpha(u+1)^p\, e^{\frac{\gamma v}{1+v}} \\
&\lambda\frac{\partial v}{\partial t} - \Delta v = \alpha\beta(u+1)^p\, e^{\frac{\gamma v}{1+v}}
\end{aligned}
\quad \text{in } \Omega\times(0,\infty)\,,
$$

$$
\begin{aligned}
&\frac{\partial u}{\partial n} + \nu u = 0 \\
&\frac{\partial v}{\partial n} + \mu v = 0
\end{aligned}
\quad \text{on } \partial\Omega\times(0,\infty)\,,
$$

$$
\begin{aligned}
u(.,0) &= u_0 \\
v(.,0) &= v_0
\end{aligned}
\quad \text{on } \overline{\Omega}\,,
$$

where $\Omega$ is a smoothly bounded domain in $\mathbb{R}^3$. These equations are derived in the book by R. ARIS [3] and describe a single, irreversible, nonisothermic, p-th order chemical reaction in a permeable catalyst whose shape is described by $\Omega$. Here $u+1$ denotes the dimensionless concentration of the reactant and $v+1$ is the dimensionless temperature. The constants $\alpha$ (the Thiele number), $\beta$ (the Prater temperature), $\gamma$ (the Arrhenius number), $\lambda^{-1}$ (the Lewis number), and $\nu,\mu$ (the Biot numbers) are positive and given, and $p$ is a positive integer (cf. [3, Section 2.5.4.] ). Finally, $n$ denotes the outer normal to the boundary $\partial\Omega$ of $\Omega$.

The system (1) is of considerable importance in the theory of chemical reactions. However, so far there seem to be no general existence, uniqueness, and stability results, although many particular cases and approximations have been considered. Indeed, large parts of ARIS' book are devoted to special cases. For example, there are studied isothermic reactions (which means that the problem can be reduced to the study of a single equation), or the case where diffusion can be neglected (in which case the system is reduced to the much simpler case of ordinary differential equations).

A great number of the investigations are devoted to the special case that $\lambda = 1$ and $\nu = \mu$. In this case, the new dependent variable $w := \beta u + v$ satisfies the homogeneous linear equations

$$\frac{\partial w}{\partial t} - \Delta w = 0 \qquad \text{in} \quad \Omega \times (0,\infty) \,,$$
$$\frac{\partial w}{\partial n} + \nu w = 0 \qquad \text{on} \quad \partial\Omega \times (0,\infty) \,,$$

which possess exactly one steady state, namely $w = 0$. Hence it follows for every steady state $(u,v)$ of (1), that $u = -\beta^{-1}v$. By inserting this value in the second equation, the problem of existence, uniqueness, and stability of steady states for (1) is reduced to a single nonlinear elliptic equation of the form

$$-\Delta v = \alpha\beta^{1-p}(\beta - v)^p e^{\frac{\gamma v}{1+v}} \qquad \text{in} \quad \Omega \,,$$
$$\frac{\partial v}{\partial n} + \nu v = 0 \qquad \text{on} \quad \partial\Omega \,.$$

For equations of this type much information has been obtained within the last few years (cf. [1], [3, Chapter 6]).

The above approach depends, of course, heavily on the assumptions $\lambda = 1$ and $\nu = \mu$. However ARIS [3, Section 2.7] points out that in concrete situations $\lambda$ can take values between $10^{-2}$ and $10^3$ and that in great many cases $\nu/\mu$ is of the order of $10^2$. This shows that the above mathematical approximations are in many cases rather unrealistic.

In the following theorem we show that none of the above assumptions is necessary to prove that the system (1) has a unique global solution. For a precise statement we introduce the following notation:
$C_B^2(\overline{\Omega}, \mathbb{R}^2) := \{u, v \in C^2(\overline{\Omega}) \mid \frac{\partial u}{\partial n} + \nu u = 0 \text{ and } \frac{\partial v}{\partial n} + \mu v = 0 \text{ on } \partial\Omega\}$.
Then the following result is true:

*Theorem 1: For every* $(u_0, v_0) \in C_B^2(\overline{\Omega}, \mathbb{R}^2)$ *satisfying* $-1 \le u_0 \le 0$ *and* $v_0 \ge -1$, *there exists exactly one (classical) solution* $(u,v)$ *of problem* (1). *Moreover,* $-1 \le u(.,t) \le o$ *and* $v(.,t) \ge -1$ *for all* $t \ge o$ .

We emphasize the fact that there are no restrictions whatsoever for the various constants in (1), besides of the positivity requirement. Moreover the restriction $-1 \le u_0 \le o$ (which is to be understood pointwise) is quite natural since in the derivation of system (1), the dimensionless concentration $u_0 + 1$ is normalized as to have its maximal value less or equal to $1$ . Similarly, the dimensionless temperature is to be nonnegative, which is reflected in the inequality $v_0 + 1 \ge o$ .

In fact, it is possible to improve the assertion of Theorem 1 somewhat. Namely, it can be shown that, for every initial temperature $v_0$ , there is an upper bound $w(v_0)$ such that $o \le v(.,t) + 1 \le w(v_0)$ for all $t \ge o$ .

The *proof* of Theorem 1 is based on the theory of semilinear evolution equations in Banach spaces as developed by Sobolevskii, Friedman, and others, as well as on the "classical" $C_\alpha$- and $L_p$-theories for linear parabolic equations, developed by Ladyzenskaja, Solonnikov, Ural'seva and others. Using these results and some tricks, the problem is reduced to the problem of finding a bounded invariant set $\mathbb{M}$ in $C(\bar{\Omega},\mathbb{R}^2)$. That is, we have to find a bounded subset $\mathbb{M}$ of $C(\bar{\Omega},\mathbb{R}^2)$ such that, *a priori,* every solution $(u,v)$ of (1) satisfies $(u(.,t),v(.,t)) \in \mathbb{M}$ for all $t > o$, provided $(u_0,v_0) \in \mathbb{M} \cap C^2_B(\bar{\Omega},\mathbb{R}^2)$ . By using the special structure of the nonlinearities and maximum principle arguments, it can be shown that such an invariant bounded set $\mathbb{M}$ exists.

By the general existence theory for semilinear evolution equations of parabolic type, it follows now that problem (1) defines a nonlinear semigroup $\{S(t),t \ge o\}$ on

$$\mathbb{M}_2 := \{(u,v) \in C^2_B(\bar{\Omega},\mathbb{R}^2) \mid (u(x),v(x)) \in \mathbb{M} \text{ for all } x \in \bar{\Omega}\} .$$

More precisely, for every $t \ge o$ , $S(t) : \mathbb{M}_2 \to \mathbb{M}_2$ is a continuous (nonlinear) map and

(i) $S(o) = id$ ,

(ii) $S(t+\tau) = S(t) \circ S(\tau)$ for $t,\tau \ge o$ .

Here $S(t)(u_0,v_0)$ denotes simply the solution $(u(.,t),v(.,t))$ at time

$t$ of problem (1). By means of regularity arguments it can be shown that the semigroup $\{S(t) \mid t \geq o\}$ has an important smoothing property, namely $S(t)(\mathbb{M}_2)$ is *relatively compact* in $\mathbb{M}_2$ (with the topology induced by $C^2(\overline{\Omega},\mathbb{R}^2)$) for every $t > o$. Hence, by Schauder's fixed point theorem, it follows that

$$\mathcal{F}(t) := \{m \in \mathbb{M}_2 \mid S(t)m = m\} \neq \emptyset$$

for every $t > o$.

Let $t_1,\ldots,t_m$ be fixed positive numbers and suppose that $t > o$ is a common divisor of all of them, that is, there exist $k_1,\ldots,k_m \in \mathbb{N}$ such that $t_j = k_j t$ for $j = 1,\ldots,m$. Then, for every $m \in \mathcal{F}(t)$,

$$S(t_j)m = S(k_j t)m = [S(t)]^{k}m = m \quad ,$$

which shows that

$$\emptyset \neq \mathcal{F}(t) \subset \bigcap_{j=1}^{m} \mathcal{F}(t_j) \quad .$$

This implies that the family $\{\mathcal{F}(t) \mid t \in \mathbb{Q}_+\}$ has the finite intersection property, and, consequently,

$$\bigcap_{t \in \mathbb{Q}_+} \mathcal{F}(t) \neq \emptyset$$

by the compactness of $\mathcal{F}(t)$, $t > o$. Hence there exists an element $m \in \mathbb{M}_2$ such that

$$S(t)m = m \quad \text{for all} \quad t \in \mathbb{Q}_+ \quad .$$

Finally, by using the continuity of the function $S(.)m$ (in an appropriate topology) and the density of $\mathbb{Q}_+$ in $\mathbb{R}_+$, it follows that

$$S(t)m = m \quad \text{for all} \quad t \geq o \; .$$

Thus $m$ is a rest point of the flow defined by (1), that is, a stationary solution of (1). By this way we obtain the following

*__Theorem 2:__ There exists at least one stationary state* $(u^*,v^*)$ *of problem* (1) *satisfying* $-1 \leq u^* \leq o$ , $v^* \geq -1$ .

By using the existence of a stationary state, the theory of analytic semigroups, and Gronwall type inequalities, it can be shown that, for sufficiently small Thiele numbers $\alpha$, the stationary state is unique and globally asymptotically stable. More precisely, the following result is true:

*__Theorem 3:__ There exists a positive constant* $\alpha_0$ *such that problem* (1) *has for every* $\alpha \in (o,\alpha_0)$ *a unique stationary state* $(u^*,v^*)$ *satis-*

*fying* $-1 \le u^* \le 0$ , $v^* \ge -1$ . *Moreover, for every* $(u_0, v_0) \in C_B^2(\bar{\Omega}, \mathbb{R}^2)$ *satisfying* $-1 \le u_0 \le 0$ *and* $v_0 \ge -1$ *, the unique solution* $(u,v)$ *of problem* (1) *satisfies*

$$\lim_{t \to \infty} \|(u(.,t),v(.,t))-(u^*,v^*)\|_{C^1(\bar{\Omega},\mathbb{R}^2)} = 0 .$$

For more details, proofs, and generalizations we refer to [2].

## References

[1] H. AMANN: Fixed point equations and nonlinear eigenvalue problems in ordered Banach spaces. SIAM Review 18 (1976), 620-709.

[2] H. AMANN: Existence and stability of solutions for semi-linear parabolic systems, and applications to some diffusion-reaction equations. Proc. Roy. Soc. Edinburgh, Series A, in press.

[3] R. ARIS: The Mathematical Theory of Diffusion and Reaction in Permeable Catalysts. Clarendon Press, Oxford 1975.

*Differential Equations and Applications*
*W. Eckhaus and E.M. de Jager (eds.)*
*©North-Holland Publishing Company (1978)*

# PERIODIC SOLUTIONS OF SEMILINEAR ELLIPTIC EQUATIONS IN A STRIP

Klaus Kirchgässner
Math.Institut A
Universität Stuttgart
Stuttgart, W.Germany

A parameter-dependent semilinear elliptic boundary value problem is considered in a strip. It is shown for some parameter interval that, if the nonlinearity satisfies certain symmetry conditions, all "small" solutions are periodic in the unbounded variable. The method described is generalisable to higher order elliptic equations.

## INTRODUCTION

The following boundary value problem is considered

$$(1)\qquad \begin{aligned} &\Delta u + \lambda u + f(u, \partial_x u, \partial_y u) = 0 \\ &u(0,y) = u(1,y) = 0\ , \quad (x,y) \in \Omega = (0,1) \times R \end{aligned}$$

Here, $\Delta$ denotes the two-dimensional Laplacean, $\lambda$ a real parameter, and f a $C^2$- function of its arguments which is horizontal at $\underline{0}$. One might consider (1) as a model equation for the Navier-Stokes system if f is chosen to be $u\partial_y u$, or one could consider (1) as the stationary part of a reaction-diffusion equation. It is quite easy to prove, namely by restricting the consideration to y-periodic solutions of a given period, that (1) has a continuum of bifurcation points, provided f satisfies certain symmetry conditions. Let us assume for a moment, that (1) is the stationary part of some evolution equation in time and that selection of certain solutions of (1) is understood through their stability- and instability properties. Then the question which pattern is selected requires two main answers, namely the determination of all solutions and the study of their stability.

The nonstandard aspect of this bifurcation problem is due to the fact that the differential operators in (1) and the domain $\Omega$ are invariant under translations in y-direction.

Hence, periodicity in y with any period is an additional condition consistent with (1). In this respect (1) is the simplest nontrivial model for problems in hydrodynamical stability, such as the Bénard- problem, or problems in phase transitions which are invariant under the Euclidean group E(2) of the plane [ 4] , [ 7], [ 8]. Recently all solutions with certain symmetry properties have been determined successfully by group-theoretic methods [ 8]. However, the basic assumption of periodicity, though questioned, has never been justified mathematically.

In this contribution we give a partial answer to the question raised above for the equation (1). We classify, for values of $\lambda$ less than $4\pi^2$, all solutions of (1) in a suitable neighborhood of 0. For $\lambda$ less than $\pi^2$ the trivial solution $u = 0$ is locally unique, for $\lambda$ between $\pi^2$ and $4\pi^2$ all"small" solutions are periodic in y, if f has certain symmetry properties. Existence of "singular" solutions can be shown if a condition for the geometry of the bifurcation picture is met. For $\lambda$ greater than $4\pi^2$ the problem is still unsolved.

Since the stability question has been answered elsewhere we omit it here [ 3]. While the proof of Theorem 2 appears elsewhere, we present a new proof of Theorem 1 which, in contrast to that in [ 5], can be generalised to higher order elliptic equations with coefficients depending smoothly on x. I am indebted to Dr.J. Scheurle for many helpful discussions.

## UNIQUENESS

Let $\mathcal{S}'$ denote the space of tempered distributions on $\Omega$, define the weight function $g_k(y) = (1 + y^2)^{k/2}$ for any natural number $k \in N_o$, and consider the real Hilbert spaces

$$H_k^o = \{ u \in L_{2,loc}(\bar{\Omega}) \ / \ g_k^{-1} u \in L_2(\Omega)\}$$

$$\mathring{H}_k^2 = \{ u \in H_k^o \ / \ g_k^{-1} \Delta u \in L_2(\Omega) \}$$

with the inner products

$$(u,v)_{o,k} = \int_\Omega g_k^{-2} uv \, dxdy \quad , \quad (u,v)_{2,k} = (u,v)_{o,k} + (\Delta u, \Delta v)_{o,k}$$

Moreover we need

$$H^{-k} = \{u \in \mathcal{S}' \ / \ u = \Sigma \, \partial_y^\beta g_\beta, \ o \leq \beta \leq k, \ g_\beta \in L_2(\Omega) \}$$

with the norm

$$\|u\|_{-k} = \inf \{ \Sigma_\beta \|g_\beta\|_{o,o}^2 \}^{1/2}$$

where the infimum is computed over all representations of u of the form $u = \Sigma\partial_y^\beta g_\beta$, $g_\beta \in L_2(\Omega)$. It is well known (c.f.[9]) that the Fourier transform $F$ with respect to y defines an isomorphism from $H^{-k}$ onto $H^o{}_k$.

LEMMA 1

For $\lambda < \pi^2$ and for every $k \in N_o$, the continuous operator

(2) $A_\lambda = (\Delta + \lambda) \quad : \quad H_k^2 \to H_k^o$

has a continuous inverse.

Proof: Consider the operator $\mathcal{L} = \partial_{xx}^2 + (\lambda - \eta^2)$ and the corresponding Green's function $G(x,\xi;\eta)$ for o-boundary conditions at x=0 and x=1. For $g \in L_2(\Omega)$ resp. $\mathcal{T}(\Omega)$, the function

(3) $v(x,\eta) = \int_0^1 G(x,\xi;\eta)g(\xi,\eta)\, d\xi$

lies in $L_2(\Omega)$ resp. $\mathcal{T}(\Omega)$ and vanishes for x=o and 1. ( v has enough regularity in x to define the trace.) Hence $\mathcal{L}|\mathcal{T}$ is onto $\mathcal{T}$ implying $\mathcal{L}|\mathcal{T}'$ to be injective. The restriction $\mathcal{L}|H^{-k} = \mathcal{L}_{-k}$ is onto $H^{-k}$ as can be seen by differentiation of $\mathcal{L}v = g$, $g,v \in L_2$, and by the definition of $H^{-k}$. Since $H^{-k} \subset \mathcal{T}'$, $\mathcal{L}_{-k}$ is invertible and its inverse $\mathcal{L}_{-k}^{-1}$ is continuous. Moreover

(4) $(\mathcal{L}_{-k}^{-1} g)(x_o, \cdot\,) = 0 \quad \text{for} \quad x_o = 0 \quad \text{and} \quad x_o = 1$

Now consider

(5) $A_\lambda\, u = f \quad , \quad f \in H_k^o$

Set $u = Fv$, $f = Fg$, $v = \mathcal{L}_{-k}^{-1} g$, then the sequence

$$f \xrightarrow{F^{-1}} g \xrightarrow{\mathcal{L}_{-k}^{-1}} v \xrightarrow{F} u$$

yields that $A_\lambda^{-1}$ exists and is continuous in $H_k^o$. The equations (4) and (5) imply $u \in H_k^2$ and thus the assertion.

In order to formulate the uniqueness result we cover $\Omega$ with a sequence of compacta

$$K_\ell = [0,1] \times [(\ell-1), \ell\,] \quad , \quad \ell \in Z$$

$H^m(K_\ell)$ denotes the usual Sobolev-space of order m.

THEOREM 1

Let be $f \in C^2(R^3,R)$, assume $f(0) = 0$, $\nabla f(0) = 0$. If $\lambda < \pi^2$ then there exists an $\varepsilon > 0$ such that, for any two solutions $u, \tilde{u} \in H^2_{loc}(\bar{\Omega})$ of (1) satisfying

$$\sup_j \|u\|_{H^2(K_j)} < \varepsilon , \quad \sup_j \|\tilde{u}\|_{H^2(K_j)} < \varepsilon$$

it follows $u = \tilde{u}$.

Proof: Define $f^j(x,y) = f(x,y+j)$ and suppose

$$\sup_j \|f^j\|_{H^0(K_1)} < \infty$$

for some $f \in H^0_{loc}(\bar{\Omega})$. Then we have

$$\begin{aligned}\|A_\lambda^{-1} f^j\|^2_{H^2(K_1)} &\leq c_1 \|A_\lambda^{-1} f^j\|^2_{2,2} \leq c_2 \|f^j\|^2_{0,2} \\ &= c_2 \sum_\ell \|g_2^{-1} f^j\|^2_{H^0(K_\ell)} \\ &\leq c_3 \sup_j \|f^j\|_{H^0(K_1)} \sum_\ell \frac{1}{1+(\ell-1)^2}\end{aligned}$$

and therefore

$$(6) \qquad \sup_j \|A_\lambda^{-1} f^j\|_{H^2(K_1)} < \gamma \sup_j \|f^j\|_{H^0(K_1)}$$

holds. Define $T : u \to f(u, \partial_x u, \partial_y u)$ which, in view of the smoothness assumptions on $f$, is a continuous map from $H^2(K)$ into $H^0(K)$ for every compact set $K \subset \Omega$. Moreover, for every $\rho > 0$, there exists a $\delta > 0$ such that

$$\|T(u) - T(\tilde{u})\|_{H^0(K_1)} < \rho \|u - \tilde{u}\|_{H^2(K_1)}$$

if $\|u\|_{H^2(K_1)}$, $\|\tilde{u}\|_{H^2(K_1)}$ are less than $\delta$. Now let $u$ and $\tilde{u}$ be solutions of (1); choose $\gamma\rho < 1$ and $\varepsilon$ to be a corresponding $\delta$. Using (6) we obtain

$$\begin{aligned}\sup_j \|u^j - \tilde{u}^j\|_{H^2(K_1)} &= \sup_j \|A_\lambda^{-1}(T(u^j) - T(\tilde{u}^j))\|_{H^2(K_1)} \\ &< \gamma\rho \sup_j \|u^j - \tilde{u}^j\|_{H^2(K_1)}\end{aligned}$$

which implies the assertion.

The method of proof can be generalized immediately to higher order uniformly elliptic operators $L + \lambda = A$ in the strip $\Omega$ with coefficients depending smoothly on the bounded variable $x$ together with homogeneous Dirichlet boundary conditions. If $\lambda$ is such that $\ker(L+\lambda) = \{0\}$ then $u = 0$ is an isolated solution among all solutions with uniformly small $H^{2m}(K_j)$-

norm - 2m being the order of L -. As an example consider the two-dimensional boundary value problem describing all time-independent perturbations of plane Poiseuille flow [2]

$$(7) \qquad \Delta^2\psi + \lambda(-2\partial_x\psi - u_o\partial_x(\Delta\psi)) + \lambda\partial_x\psi\partial_y(\Delta\psi) - \partial_y\psi\partial_x(\Delta\psi))$$

$$\psi = \partial_y\psi = 0 \quad \text{on } \partial\Omega$$

where $u_o(x) = x(1-x)$, $\psi$ the stream function and $\lambda$ the Reynolds number. Since (7) falls into the framework of this analysis we conclude that plane Poiseuille flow is an isolated solution of the Navier-Stokes system in the sense described above, as long as the kernel of the derivative of (7) at 0 is$\{0\}$. Detailed proofs of this generalization cam be found in a forthcoming paper.

There is an extension of Theorem 1 for equation (1) beyond $\lambda = \pi^2$. In a previous paper [5] it was shown that uniqueness modulo $\ker(\Delta+\lambda)$ holds locally for all $\lambda \in R$. To be precise, let be

$$X^s = \bigcup_{k=1}^{\infty} H_k^2 \quad , \quad s = 0 \text{ or } 2$$

the inductive limit of the spaces $H_k^s$. Then $A_\lambda : \hat{X}^2 \to X^0$ is always surjective. If $\lambda \in (n^2\pi^2, (n+1)^2\pi^2)$, the kernel of $A_\lambda$ is spanned by the functions

$$\varphi_{2j-1} = \sin j\pi x \cos \omega_j y \quad , \quad \omega_j = \{\lambda - j^2\pi^2\}^{1/2}$$

$$\varphi_{2j} = \sin j\pi x \sin \omega_j y \quad , \quad j = 1,\dots,n$$

We define the Fourier coefficients of u and a projector as follows

$$u_j(y) = \sqrt{2}\int_0^1 \sin j\pi x \, u(x,y)\, dx$$

$$P_n u = \sum_1^n (u_\nu(0)\varphi_{2\nu-1} + \frac{1}{\omega_\nu} u'_\nu(0)\varphi_{2\nu})$$

<u>THEOREM 2</u>

Assume that $\lambda \in (n^2\pi^2, (n+1)^2\pi^2)$, $n \in N$, then there exists a positive number $\varepsilon$ such that, given any two solutions $u$ and $\tilde{u}$ of (1) satisfying

$$\sup_j \|u\|_{H^2(K_j)} < \varepsilon \quad , \quad \sup_j \|\tilde{u}\|_{H^2(K_j)} < \varepsilon$$

$$P_n u = P_n u$$

the two solutions coincide.

For the proof see [5]. For a complete description of all small solutions of (1) it suffices to show that"above" every

$\varphi \in \ker(\Delta + \lambda)$ there exists at least one solution.

EXISTENCE

What we have said about the kernel of $\Delta + \lambda$ suggests that, for $\lambda \in (\pi^2, 4\pi^2)$, and under suitable assumptions on f, all "small" solutions of (1) should be periodic in y. That additional conditions on f are necessary is shown by the example $f = (\partial_y u)^3$, for which no y-periodic solution exists, except $u = 0$. Let us therefore assume

(8)
(a) $f(u,p,-q) = f(u,p,q)$
(b) $f(u,p,-q) = -f(u,p,q)$ and $f(-u,-p,-q) = f(u,p,q)$

Consider the case 8a), set $u(x,y) = v(x,\omega y)$ and determine, for fixed $\lambda \in (\pi^2, 4\pi^2)$, nontrivial solutions of

(9) $B(\omega)v + \lambda v + f(v, \partial_x v, \omega\partial_z v) = 0$

$v(0,z) = v(1,z) = 0$ , $v(x,\cdot)$ $2\pi$ - periodic

where $B = \partial^2_{xx} + \omega^2 \partial^2_{zz}$

We consider (9) as a bifurcation problem near $\omega = \omega_1 = (\lambda - \pi^2)^{1/2}$ and $v = 0$. If we impose the further requirement that v should be even in z, the operator $B(\omega)$, being selfadjoint in $L_2((0,1)x(0,2\pi))$, has a 1-dimensional kernel for $\omega = \omega_1$. Hence, by a well known theorem [1] , $\omega = \omega_1$, $v = 0$ is a bifurcation point.

Since,for every solution u, $u^c(x,y) = u(x,y+c)$ is a solution as well, one obtains a two-dimensional manifold of y-periodic solutions of (1) which is modelled over $\ker(\Delta + \lambda)$ (c.f. [5]). The case 8b) can be treated similarly. Hence we have

THEOREM 3

Let be $\lambda \in (\pi^2, 4\pi^2)$ and assume one of the conditions (8) to hold. Then there exists a positive number $\varepsilon_o$ such that, if $|P_1 u| < \varepsilon_o$, and if u is a solution of (1), then u is periodic in y.

Conversely, for every $\varepsilon \in (0, \varepsilon_o]$ a solution u of (1) exists satisfying $|P_1 u| = \varepsilon$.

The case $\lambda > 4\pi^2$ is much more difficult to solve. The kernel of ( $\Delta + \lambda$ ) consists of quasiperiodic functions, i.e. functions of the form $u(x,y) = v\ (x, \omega_1 y, \ldots, \omega_n y)$, where $v(x, z_1, \ldots, z_n)$ is $2\pi$ - periodic in every $z_j$. The study of the full nonlinear equation leads to problems of small divisors (see [3]). Nothing is known about existence.

The point $\lambda = \pi^2$ may be a point of bifurcation for singular solutions, i.e. functions whose first Fourier component is either constant or nonperiodic. Let us consider the set $S_\lambda = \{u \in X^2 \ / \ u$ belongs to the component of solutions bifurcating at $(0,\omega_1)\}$. If $\cup S_\lambda$, $\lambda \in (\pi^2-\delta,\pi^2+\delta)$ for some $\delta > 0$, is confined to some domain $D \subset (\pi^2,\infty) \times X^2$, and if $D \cap (\{\omega\} \times X^2) = D_\omega$ shrinks to $\{(\omega_1,0)\}$ as $\lambda$ approaches $\pi^2$ from above, then there exist, for $\lambda \in (\pi^2,\pi^2+\delta)$, nontrivial solutions of arbitrary large irreducible periods. They converge with increasing period towards a singular solution in $X^2$. If this geometric condition is violated, singular solutions may not exist. The proof is a consequence of Theorem 2 and of the global bifurcation result of Rabinowitz [6].

## REFERENCES


[1] Crandall, M.G. and Rabinowitz, P.H., (1971), Bifurcation from simple eigenvalues, J. Functional Anal., 8, pp. 321-340.

[2] Joseph, D.D., (1976), Stability of fluid motions, I, II, Springer-Verlag, Berlin.

[3] Kirchgässner, K., (1977), Preference in pattern and cellular bifurcation in fluid dynamics, in Applic. of ifurcation theory, P.Rabinowitz ed., Academic Press, pp.149-173.

[4] Kirchgässner, K.and Kielhöfer, H., (1972), Stability and bifurcation in fluid mechanics, Rocky Mountain J. Math., 3, pp. 275-318.

[5] Kirchgässner, K. and Scheurle, J., (1977), On the bounded solutions of a semilinear elliptic equation in a strip, manuscript, to appear.

[6] Rabinowitz, P.H., (1971), Some global results for nonlinear eigenvalue problems, J. Functional Anal., 7, pp. 487-513.

[7] Raveché, H.J. and Stuart, C.A., (1976), Bifurcation of solutions with crystalline symmetry, J. Math. Phys., 17, pp. 1949-1953.

[8] Sattinger, D.H., (1977), Group representation theory, bifurcation theory, and pattern formation, (1977), J. Functional Anal., to appear.

[9] Trèves, F., (1967), Topological vector spaces, distributions and kernels, Academic Press, New York.

*Differential Equations and Applications*
*W. Eckhaus and E.M. de Jager (eds.)*
*©North-Holland Publishing Company (1978)*

# TIME PERIODIC SOLUTIONS OF A SEMILINEAR WAVE EQUATION

Paul H. Rabinowitz*
Mathematics Department
University of Wisconsin
Madison, Wisconsin 53706

The purpose of this talk is to describe some recent work on the existence of time-periodic solutions of a semilinear wave equation. Our simplest result is for the problem of finding time-periodic solutions of

$$(1) \qquad \begin{cases} u_{tt} - u_{xx} + f(u) = 0, & 0 < x < \pi, \ t \in \mathbb{R} \\ u(0,t) = 0 = u(\pi, t) \end{cases}$$

where $f(0) = 0$. Hence $u \equiv 0$ is such a solution and we seek a nontrivial time-periodic solution. Note that the period is not known a priori. Our main result for (1) is:

<u>Theorem 2:</u> Suppose $f$ satisfies

($f_1$) $f \in C^2(\mathbb{R}, \mathbb{R})$,

($f_2$) $f$ is strictly monotone increasing,

($f_3$) $f$ is superlinear at $0$ and $\infty$, i.e.

(i) $f(z) = o(|z|)$ at $z = 0$,

(ii) $F(z) = \int_0^z f(s)ds \leq \Theta z f(z)$ for large $|z|$ where $\Theta \in (0, \frac{1}{2})$.

Then for any period $\tau$ which is a rational multiple of $\pi$, (1) possesses a nontrivial $C^2$ time-periodic solution.

The proof of Theorem 2 relies on minimax existence techniques from the calculus of variations and regularity arguments from the theory of elliptic partial differential equations. It can be found in [1].

---

*This research was sponsored in part by the Office of Naval Research under Contract N00014-76-C-0300 and in part by the Army Research Office under Contract DAAG29-75-C-0024. Reproduction in whole or in part is permitted for any purpose of the U. S. Government.

A few remarks are in order about the statement of the theorem and some generalizations. First of all, the function $f$ is called superlinear at $\infty$ since $(f_3)$ (ii) implies that

(3) $$\frac{1}{\Theta z} \le \frac{F'}{F}$$

or

(4) $$F(z) \ge c|z|^{\frac{1}{\Theta}}, \quad |f(z)| \ge c\Theta^{-1}|z|^{\frac{1}{\Theta}-1}$$

for large $|z|$. Secondly, the reason for the rationality condition on $\tau$ is a technical one. For $\tau$ rational, the spectrum of the wave operator, $\Box$, is discrete and 0 is an isolated point in the spectrum of $\Box$. This enables us to invert $\Box$ on the orthogonal complement of the null space of $\Box$ and in fact we gain one derivative (in either the maximum or $L^2$ norm) in so doing. On the other hand if $\tau$ is irrational, the spectrum of $\Box$ is dense with 0 as an accumulation point. point.

There are several ways in which Theorem 2 can be extended: (i) $f(z)$ can be replaced by $az + f(z)$ with $a > 0$; (ii) $f$ can be permitted to depend on $x$ and $t$, the latter in a periodic fashion (and solutions having the same period are obtained, again provided that the period satisfies our rationality condition); (iii) if $f$ is merely assumed to be continuous and strictly is removed from $(f_2)$, there still exist weak solutions of (1) in $L^2 \cap L^\infty$. Details can be found in [1].

## REFERENCES

[1] Rabinowitz, P. H., Free vibrations for a semilinear wave equation, to appear Comm. Pure Appl. Math.

*Differential Equations and Applications*
*W. Eckhaus and E.M. de Jager (eds.)*
*©North-Holland Publishing Company (1978)*

# TWO-SIDED BOUNDS AND NORM BOUNDS FOR SYSTEMS OF NONLINEAR DIFFERENTIAL EQUATIONS

Johann Schröder *)
Mathematisches Institut
Universität Köln

The paper is concerned with estimates for solutions of systems of second order differential equations. Pointwise two-sided bounds and pointwise norm bounds are considered. In both cases a priori estimates are derived as well as results on existence and estimation.

## 1. Introduction

For solutions $u^*$ of certain systems of ordinary or partial differential equations $Mu = o$ of the second order, we shall report on *pointwise two-sided estimates*

$$\varphi(x) \leq u^*(x) \leq \psi(x) \ , \tag{1.1}$$

where $\leq$ is interpreted to hold componentwise, and on *pointwise norm estimates*

$$\|u^*(x)\| \leq \Psi(x) \ , \tag{1.2}$$

where $\| \ \|$ denotes a certain vector norm.

Section 2 yields a priori estimates; Section 3 is concerned with existence statements and estimations. A comparison of these two types of estimates (1.1) and (1.2) is provided in Section 4. Two-sided estimates, for example, have the advantage of yielding componentwise bounds. On the other hand, there are advantages to norm estimates, if the solution $u^*$ has an "oscillatory behavior" or if the system $Mu = o$ is strongly coupled.

For ordinary differential equations proofs can be found in [5],[6],[7]. The results for partial differential equations can be obtained by essential-

*) The research reported herein has been sponsered in part by the European Research Office (U.S. Army).

ly the same methods. Only the existence proofs require some additional considerations here. For these reasons no proofs will be given in Section 2, and the proofs in Section 3 will only be sketched. (See the above papers for a discussion of the results, for generalizations and applications.)

A comment on our notation. For $G \subset \mathbb{R}^m$, $\mathbb{R}^n(G)$ denotes the set of functions $u : G \to \mathbb{R}^n$ ; we write $u = (u_i)$ with $u_i \in \mathbb{R}(G) := \mathbb{R}^1(G)$ $(i = 1,2,\dots,n)$ . $C^n_k(G)$ is the subset of all $u \in \mathbb{R}^n(G)$ such that $u$ is k times continuously differentiable on $G$ (provided this term is meaningful); $u_x$ is the $n \times m$ matrix with components $\partial u_j/\partial x_k$ $(j = 1,2,\dots,n;\ k = 1,2,\dots,m)$. $\mathbb{R}^{n,m}$ is defined to be the set of real $n \times m$ matrices.

The sign $\leq$ denotes the natural (point- and componentwise) order relation, while $\underset{=}{\prec}$ denotes the strict natural order relation. Thus, if $G$ is compact and $u \in C^n_0(G)$ , then $u \succ o$ if and only if $u_i(x) > 0$ for all $x \in G$ and $i=1,2,\dots,n$ . Null-functions and null-vectors are denoted by $o$ .

Throughout this paper, $\langle\ ,\ \rangle$ denotes an inner product in $\mathbb{R}^n$ and $\|y\| = \langle y,y\rangle^{\frac{1}{2}}$ .

## 2. A priori estimates

Suppose that $\Omega$ is a bounded domain in $\mathbb{R}^m$ with boundary $\partial\Omega$ , $\bar{\Omega} = \Omega \cup \partial\Omega$ , $R = C^n_0(\bar{\Omega}) \cap C^n_2(\Omega)$ , $S = \mathbb{R}^n(\bar{\Omega})$, $R_0 = C_0(\bar{\Omega}) \cap C_2(\Omega)$ , and that $M : R \to S$ is an operator of the form

$$Mu(x) = \begin{cases} \mathcal{L}[u](x) + f(x,u(x),u_x(x)) & \text{for } x \in \Omega \\ u(x) & \text{for } x \in \partial\Omega , \end{cases}$$

where the quantities have the following properties. $\mathcal{L}[u]$ has the n components $(\mathcal{L}[u])_i = L_i[u_i]$ with $L_i[\varphi]$ defined for $\varphi \in R_0$ by

$$L_i[\varphi](x) = -\sum_{j,k=1}^{m} a^i_{jk}(x)\frac{\partial^2\varphi(x)}{\partial x_j \partial x_k} + \sum_{j=1}^{m} b^i_j(x)\frac{\partial\varphi(x)}{\partial x_j}$$

and $a^i_{jk}$ , $b^i_j$ in $\mathbb{R}(\Omega)$ such that each $m \times m$ matrix $(a^i_{jk}(x))$ $(i=1,2,\dots,n)$ is positive semi-definite. $f$ is a function such that $f(x,y,P) \in \mathbb{R}^n$ for $x \in \Omega$ , $y \in \mathbb{R}^n$ , $P \in \mathbb{R}^{n,m}$ .

In this section, let $v$ denote a fixed (but unknown) function in $R$ . We shall derive estimates for $v$ using properties of $Mv$ (*Range-Domain implications.*

In Theorem 1 we shall make the additional assumption, that the operator $M$ is *weakly coupled*; that is, for each $i \in \{1,2,\ldots,n\}$ the function $f_i(x,y,P)$ depends only on $x,y$ and the i-th row of $P$. Then the i-th component $f_i(x,u,u_x)$ does not contain any derivatives of $u_k$ with $k \neq i$. (This condition on $M$ can be weakened slightly. See Section 4.)

To formulate Theorem 1 we introduce a function $F : \Omega \times \mathbb{R}^n \times \mathbb{R}^n \times \mathbb{R}^{n,m} \to \mathbb{R}^n$ with components $F_i$ defined by

$$F_i(x,y,\eta,P) = f_i(x,y^{(i)},P) \ , \quad y_i^{(i)} = y_i \ , \quad y_k^{(i)} = \eta_k \quad \text{for } k \neq i \ .$$

The variable $x$ will be omitted at some places; for example, we write $F_i(x,\psi,h,\psi_x)$ for $F_i(x,\psi(x),h,\psi_x(x))$, where $\psi$ is a function of $x$ and $h$ is a constant vector.

Theorem 1. *Let* $M$ *be weakly coupled and* $v \in R$. *Suppose there exist functions* $\varphi,\psi,z,\bar{z}$ *in* $R$ *such that* $\varphi \leq \psi$, $z \succ o$, $\bar{z} \succ o$ *and that for each* $\lambda > 0$ *with* $\varphi - \lambda\bar{z} \leq v \leq \psi + \lambda z$ *and each index* $i \in \{1,2,\ldots,n\}$ *the following two conditions hold.*

(i) *For each* $x \in \Omega$ *with* $(\varphi_i - \lambda\bar{z}_i - v_i)(x) = 0$, $(\varphi_i - \lambda\bar{z}_i - v_i)_x(x) = 0$, *there exists a vector* $\bar{h} \in \mathbb{R}^n$ *such that* $\varphi(x) \leq \bar{h} \leq \psi(x)$ *and*

$$L_i[\bar{z}_i](x) + \lambda^{-1}[F_i(x,\varphi,\bar{h},\varphi_x) - F_i(x,\varphi - \lambda\bar{z}_i,v,\varphi_x - \lambda\bar{z}_x)] > 0 \ .$$

(ii) *For each* $x \in \Omega$ *with* $(\psi_i + \lambda z_i - v_i)(x) = 0$, $(\psi_i + \lambda z_i - v_i)_x(x) = 0$, *there exists a vector* $h \in \mathbb{R}^n$ *such that* $\varphi(x) \leq h \leq \psi(x)$ *and*

$$L_i[z_i](x) + \lambda^{-1}[F_i(x,\psi + \lambda z,v,\psi_x + \lambda z_x) - F_i(x,\psi,h,\psi_x)] > 0 \ . \tag{2.1}$$

*Then* $\varphi \leq v \leq \psi$ *holds, if*

$$\varphi(x) \leq v(x) \leq \psi(x) \quad \textit{for} \quad x \in \partial\Omega$$

*and if*

$$\mathcal{L}[\varphi](x) + F(x,\varphi,\bar{h},\varphi_x) \leq Mv(x) \leq \mathcal{L}[\psi](x) + F(x,\psi,h,\psi_x) \tag{2.2}$$

*for all* $x \in \Omega$, $h \in \mathbb{R}^n$, $\bar{h} \in \mathbb{R}^n$ *satisfying* $\varphi(x) \leq h \leq \psi(x)$, $\varphi(x) \leq \bar{h} \leq \psi(x)$.

This result is proved by essentially the same method as Theorem 5.2 in [6].

In the next theorem we need not require that $M$ is weakly coupled. Here, however, we make the additional assumption that all operators $L_i$ are the same, that is,

$$a^i_{jk} = a_{jk} , \quad b^i_j = b_j \quad \text{for all} \quad i \in \{1,2,\ldots,n\} . \tag{2.3}$$

Then we denote by $A(x)$ the matrix $A(x) = (a_{jk}(x))$ and write $L = L_i$. Moreover, we define $tr(B) = \Sigma^n_{k=1} b_{kk}$ for $B = (b_{jk}) \in \mathbb{R}^{n,n}$.

Theorem 2. *Suppose that* (2.3) *holds and that there exist functions* $\Psi \geq o$ *and* $z \succ o$ *in* $R_o$ *such that the following inequality is satisfied:*

$$0 < L[z](x)+tr(QA(x)Q^T)z(x)+\lambda^{-1}\langle \eta, f(x,v(x),v_x(x))-f(x,\Psi(x),\eta,\eta\Psi_x(x)+\Psi(x)Q)\rangle$$

*for all* $x \in \Omega$, $\lambda \in \mathbb{R}$, $\eta \in \mathbb{R}^n$, $Q \in \mathbb{R}^{n,m}$ *with*

$$\lambda > 0, \ \|\eta\| = 1, \ \eta^T Q = o, \ v(x) = (\Psi+\lambda z)(x)\eta, \ v_x(x) = \eta(\Psi_x+\lambda z_x)(x)+(\Psi+\lambda z)(x)Q .$$

*Then* $\|v(x)\| \leq \Psi(x)$ *holds for all* $x \in \bar{\Omega}$, *if*

$$\|Mv(x)\| \leq L[\Psi](x) + tr(QA(x)Q^T)\Psi(x) + \langle \eta, f(x,\Psi(x)\eta,\eta\Psi_x(x) + \Psi(x)Q)\rangle \tag{2.4}$$

*for all* $x \in \Omega$, $\eta \in \mathbb{R}^n$, $Q \in \mathbb{R}^{n,m}$ *with*

$$\|\eta\| = 1 , \quad \eta^T Q = o , \quad o \neq v(x) = \|v(x)\|\eta .$$

This result is proved by essentially the same method as Theorem 2.2 in [7].

## 3. Existence and inclusion statements

Now we consider a boundary value problem $Mu = o$ with an operator $M$ as defined in Section 2. We assume, however, that $f$ *does not depend on the derivative* $u_x$, i.e., $f(x,y,P) = f(x,y)$. Moreover, we require that $a^i_{jk} \in C_1(\bar{\Omega})$, $b^i_j \in C_1(\bar{\Omega})$, $\partial\Omega \in C_3$, that for each $x \in \bar{\Omega}$ and each $i$ the matrix $(a^i_{jk}(x))$ is positive definite and that $f(x,y)$ is a continuously differentiable function on $\bar{\Omega} \times \mathbb{R}^n$ (these conditions may be weakened).

Theorem 3. *Suppose that functions* $\varphi,\psi \in C^n_2(\bar{\Omega})$ *exist such that* $\varphi \leq \psi$,

$$\varphi(x) \leq o \leq \psi(x) \quad \textit{for all} \quad x \in \partial\Omega$$

*and*

$$\mathcal{L}[\varphi](x) + F(x,\varphi(x),\bar{h}) \leq o \leq \mathcal{L}[\psi](x) + F(x,\psi(x),h)$$

*for all* $x \in \Omega$, $h \in \mathbb{R}^n$, $\bar{h} \in \mathbb{R}^n$ *satisfying* $\varphi(x) \leq h \leq \psi(x)$, $\varphi(x) \leq \bar{h} \leq \psi(x)$.

*Then the given problem* $Mu = o$ *has a solution* $u^* \in C_2^n(\bar{\Omega})$ *such that* $\varphi \leq u^* \leq \psi$ .

We sketch the proof of this theorem. Define $Au = \mathcal{L}[u] + cu$ and $Bu = cu^{\#} - f(x,u^{\#})$ , where $u^{\#} = \sup\{\varphi,\inf\{u,\psi\}\}$ and $c > 0$ is an (arbitrarily large) constant. Moreover, choose $p \in \mathbb{N}$ and $\alpha \in \mathbb{R}$ such that $m < p < \infty$ , $0 < \alpha < 1 - mp^{-1}$ . The set $BC^n(\bar{\Omega})$ is bounded, i.e., $\|Bu\| \leq \varkappa$ for all $u \in C_0^n(\bar{\Omega})$ with $\|u\| = \max\{|u_i(x)| : 1,2,\ldots,n;\ x \in \bar{\Omega}\}$ . For sufficiently large $c$ , one derives the following from known results: For each $u \in C_0^n(\bar{\Omega})$ there exists a uniquely determined $v = (v_i)$ such that $v_i \in H_{2,p}(\Omega) \cap \overset{o}{H}_{1,p}(\Omega)$ and $Av = Bu$ . (Compare [4], Corollary 7.4, Theorem 9.12 and Remark 9.13.) In addition, $\|v_i\|_{1+\alpha} \leq c_1|v_i|_{2,p} \leq c_2\varkappa$ . Here Hölder norms are denoted by $\|\ \|$ and Sobolev norms by $|\ |$ . (For the first inequality compare [1], Sections 9 and 10; the second inequality is a consequence of $|v_i|_{2,p} \leq c_3|(Bu)_i|_{0,p}$ . Observe that in the latter inequality no term $\|u_i\|_{0,p}$ need occur since $A$ is one-to-one.) Thus, by $v = Tu$ an operator $T$ is defined which maps $C_0^n(\bar{\Omega})$ into a relatively compact subset. Since $T$ is also continuous, Schauder's fixed point theorem yields the existence of a $u^* = (u_i^*)$ such that $u_i^* \in H_{2,p}(\Omega) \cap \overset{o}{H}_{1,p}(\Omega)$ and $u^* = Tu^*$, i.e., $Au^* = Bu^*$ . The functions $u_i^*$ belong to $C_{1+\alpha}(\bar{\Omega})$ , so that the components of $Bu^*$ are in $C_\alpha(\bar{\Omega})$ . Consequently, $u_i^* \in C_{2+\alpha}(\bar{\Omega})$ and $u(x) = o$ for $x \in \partial\Omega$ , by Schauder's existence theorem (see [1], for example).

Now we define a modified operator $M^{\#} : R \to S$ by $M^{\#}u(x) = Au(x) - Bu(x)$ for $x \in \Omega$ , $M^{\#}u(x) = u(x)$ for $x \in \partial\Omega$ and apply Theorem 1 to this operator choosing $z_i(x) \equiv \bar{z}_i(x) \equiv 1$ . Here the assumptions of Theorem 1 are easily verified. For example, inequality (2.1) (for $M^{\#}$ in place of $M$ ) is equivalent to $c\lambda + [F_i(x,\psi(x),v^{\#}(x)) - F_i(x,\psi(x),h)] > 0$ . This condition holds for $v = u^*$ and $h = v^{\#}(x)$ . Since also (2.2) is satisfied for $M^{\#}u^*$ in place of $Mv$ , we obtain $\varphi \leq u^* \leq \psi$ . Consequently, $(u^*)^{\#} = u^*$ and hence $Mu^* = M^{\#}u^* = o$ .

Without giving details, we note that a similar result can be proved by essentially the same methods, if $f$ depends also on $P$ , but $f(x,y,P)$ is bounded on the set of all $(x,y,P)$ with $x \in \bar{\Omega}$, $\varphi(x) \leq y \leq \psi(x)$, $P \in \mathbb{R}^{n,m}$.

Here $f$ may even be strongly coupled. (Now one defines $B$ on $C_1^n(\bar{\Omega})$ by $Bu = cu^{\#} - f(x,u^{\#},u_x)$). Of course the above boundedness condition is also very restrictive.

The second part of the above proof can be modified such that only Theorem 1 for $n = 1$ is used (compare the proof of Theorem 3.1 in [5]).

Theorem 4. *Suppose that* (2.3) *holds and that there exists a function* $\Psi \succ o$ *in* $C_2(\bar{\Omega})$ *which satisfies*

$$0 \leq L[\Psi](x) + \langle \eta, f(x,\Psi(x)\eta) \rangle$$

*for all* $x \in \Omega$ *and all* $\eta \in \mathbb{R}^n$ *with* $\|\eta\| = 1$ .

*Then the given problem* $Mu = o$ *has a solution* $u^* \in C_2^n(\bar{\Omega})$ *such that* $\|u^*(x)\| \leq \Psi(x)$ *for all* $x \in \bar{\Omega}$ .

The first part of the proof of this theorem uses essentially the same tools as the proof of Theorem 3 in establishing the existence of a solution $u^* \in R$ of a modified problem $M^{\#}u = o$ . Now, $M^{\#}u = Au - Bu$ with $A$ as above and $Bu(x) = cu^{\#}(x) - \Delta(x,\|u(x)\|)f(x,u^{\#}(x))$ , where $\Delta(x,t) = 1$ for $t \leq \Psi(x)$ , $\Delta(x,t) = t^{-1}\Psi(x)$ for $t > \Psi(x)$ , $u^{\#}(x) = \Delta(x,\|u(x)\|)u(x)$ .

To prove that $(u^*)^{\#} = u^*$ , one proceeds as for ordinary differential operators (Theorem 3.1 in [7]).

## 4. Comparing the methods of estimation

Results similar to those above can also be derived for certain operators related to initial value problems. To compare the estimates involving two-sided bounds with those involving norm bounds we consider first a very simple ordinary differential operator $M : C_1^n[0,\ell] \to \mathbb{R}^n[0,\ell]$ given by

$$Mu(x) = \begin{cases} u'(x) + f(x,u(x)) & \text{for } 0 < x \leq \ell \\ u(0) & \text{for } x = 0 \end{cases} \tag{4.1}$$

with continuously differentiable $f : [0,\ell] \times \mathbb{R}^n \to \mathbb{R}^n$ . Here we have the following two statements which correspond to Theorems 1 and 2.

Proposition 5. *Suppose that* $\varphi,\psi,v \in C_1^n[0,\ell]$ *with* $\varphi \leq \psi$ *satisfy for each fixed index* $i$ *the following inequalities:*

$$\varphi_i(0) \leq v_i(0) \leq \psi_i(0) ,$$

$$\varphi_i'(x) + f_i(x,\bar{h}) \leq (Mv)_i(x) \leq \psi_i'(x) + f_i(x,h) \tag{4.2}$$

*for all* $x \in (0,\ell]$ *and* $h,\bar{h} \in \mathbb{R}^n$ *such that*

$$\varphi(x) \leq \bar{h} \leq \psi(x) \ , \quad \bar{h}_i = \varphi_i(x) \ , \ \varphi(x) \leq h \leq \psi(x) \ , \ h_i = \psi_i(x) \ .$$

*Then* $\varphi \leq v \leq \psi$ .

<u>Proposition 6.</u> *Suppose that* $v \in C_1^n[0,\ell]$ *and* $\Psi \in C_1[0,\ell]$ *with* $\Psi \geq o$ *satisfy* $\|v(0)\| \leq \Psi(0)$ *and*

$$\|Mv(x)\| \leq \Psi'(x) + \langle \eta, f(x,\Psi(x)\eta)\rangle \tag{4.3}$$

*for all* $x \in (0,\ell]$ *and* $\eta \in \mathbb{R}^n$ *such that* $\|\eta\| = 1$ , $o \neq v(x) = \|v(x)\|\eta$ .

*Then* $\|v(x)\| \leq \Psi(x)$ *for all* $x \in [0,\ell]$ .

In these statements conditions analogous to those imposed on $z,\bar{z}$ in Theorem 1 and those on $z$ in Theorem 2 need not be formulated explicitly, since such conditions are always satisfied here. (See Theorem 2.8 in [7], for example.)

Existence and inclusion statements are also easily proved for initial value problems $Mu = r$ with $M$ in (4.1). Using such estimates with two-sided bounds, Marcowitz [3] developed a numerical algorithm for error estimation. (Concerning norm bounds see [7], Theorem 3.8.)

In both propositions above the bounds ($\varphi,\psi$ and $\Psi$ , respectively) are required to satisfy a *family of inequalities* with parameters $h,\bar{h} \in \mathbb{R}^n$ in Proposition 5 and a parameter $\eta \in \mathbb{R}^n$ in Proposition 6. In (4.3) values of $f(x,y)$ occur such that $y$ lies on the surface of the ball $\{y : \|y\| \leq \Psi(x)\}$, while in (4.2) values of $f(x,y)$ are used such that $y$ lies on the surface of the cube $\{y : \varphi(x) \leq y \leq \psi(x)\}$ . For a concrete problem one may try to eliminate the parameters.

For example, suppose that $\langle y,\eta\rangle = y^T\eta$ and

$$f(x,y) = Cy \quad \text{with constant} \quad C = (c_{jk}) \in \mathbb{R}^{n,n}$$

and introduce $n \times n$ matrices $D,B,C_a,C_H$ by

$$D = \mathrm{diag}(c_{kk}) \ , \ C = D - B \ , \ C_a = D - |B| \ , \ C_H = \tfrac{1}{2}(C + C^T) \ .$$

(We write $|B| = (|b_{jk}|)$ , $|u| = (|u_i|)$ .) Then, for the case $\varphi = -\psi$ , *the inequalities required in Proposition* 5 *are satisfied if*

$$|v(0)| \leq \psi(0) \quad \textit{and} \quad |Mv(x)| \leq \psi'(x) + C_a\psi(x) \quad (0 < x \leq \ell) \ .$$

Moreover, *the inequalities required in Proposition* 6 *hold if*

$$\|v(0)\| \le \Psi(0) \quad \textit{and} \quad \|Mv(x)\| \le \Psi'(x) + \sigma\Psi(x) \quad (0 < x \le \ell)$$

*with* $\sigma$ *the smallest eigenvalue of* $C_H$ .

Obviously, two-sided bounds have the advantage of yielding componentwise estimates. Moreover, here only the i-th component of $Mu$ is used in the differential inequality containing $\psi_i'$ . On the other hand, norm bounds can have advantages for functions with oscillatory behavior.

Suppose, for example, that $n = 2$ , $\langle y,\eta\rangle = y^T\eta$ , and that the matrix $C$ has eigenvalues $\mu \pm i\nu$ with $\mu > 0$ , $\nu > 0$ . Let us even assume that $C = \begin{pmatrix} \mu & \nu \\ -\nu & \mu \end{pmatrix}$ . (This form can be achieved by a real coordinate transformation.) Then $C_a$ has the eigenvalues $\mu+\nu, \mu-\nu$ , while $C_H$ has the double eigenvalue $\sigma = \mu$ . Due to this fact one can, in general, find a norm bound which has essentially the same growth behavior for $x \to \infty$ as $v$ . This is, however, not possible for two-sided bounds. (For more details, see [7].)

The inequalities for $\varphi, \psi$ or $\Psi$ required in Sections 2 and 3 for operators $M$ of the second order can be discussed in a similar way, if $f(x,y,P)$ does not depend on $P$ .

If $f(x,y,P)$ *does* depend on $P$ , the inequality (2.4) contains a further parameter $Q \in \mathbb{R}^{n,m}$ and, therefore, seems more complicated than (2.2), where no further parameter occurs. However, further matrix-valued parameters would also occur in (2.2), if we had not assumed $M$ to be weakly coupled. This assumption can be weakened, although "not very much". For example, if $f(x,y,P)$ is a linear function of $P$ , then $M$ has to be weakly coupled (in Theorem 1).

For illustration let $m = 1$ , $\langle y,\eta\rangle = y^T\eta$ and $M : C_2^n[0,1] \to C_0^n[0,1]$ be given by

$$Mu(x) = \begin{cases} -u''(x) + f(x,u'(x)) & \text{for } 0 < x < 1 \\ u(0) & \text{for } x = 0 \\ u(1) & \text{for } x = 1 , \end{cases}$$

where $f(x,p)$ need not be weakly coupled. (Now we write $p$ for the $n \times 1$ matrix $P$ .) For this operator we have to require the following conditions instead of (2.2): For each $i$ , let

$$-\varphi_i''(x) + f_i(x,\bar{q}) \le (Mv)_i(x) \le -\psi_i''(x) + f_i(x,q)$$

for all $x \in (0,1)$, $\bar{q} \in \mathbb{R}^n$, $q \in \mathbb{R}^n$ such that $\bar{q}_i = \varphi_i(x)$, $q_i = \psi_i(x)$. Obviously, if $f(x,p) = Cp$ with a constant matrix $C$, these inequalities can only hold, if $C$ is diagonal, i.e., if $f$ is weakly coupled.

Theorem 2, however, does not require this condition. Here we have to verify the inequality (2.4), which contains two parameters $\eta$ and $q$. (We now write $q$ for the $n \times 1$ matrix $Q$; then $tr(QA(x)Q^T) = \langle q,q \rangle$.) One may try to use the infimum of the right-hand side of (2.4) with respect to these parameters.

For example, if $f(x,p) = Cp$ and $C = -C^T$, then inequality (2.4) becomes equivalent to

$$\|Mv(x)\| \le -\Psi''(x) + \tfrac{1}{4}\tau\Psi(x) \quad (0 < x < 1)$$

with $\tau$ the smallest eigenvalue of $C^2$. (See [7], where more general matrices $C$ are treated.) For $\tau > -4\pi^2$ a function of the form $\Psi(x) = \alpha\cos(\pi - \varepsilon)(x - \tfrac{1}{2})$ satisfies the above inequality as well as $\|v(0)\| \le \Psi(0)$.

## 5. References


1. A.Friedman (1969): Partial Differential Equations. Holt, Rinehart and Winston, Inc.

2. O.A. Ladyzhenskaya and N.N. Ural'tseva (1968): Linear and quasilinear elliptic equations. Acad. Press, New York and London.

3. U. Marcowitz (1975): Fehlerabschätzung bei Anfangswertaufgaben für Systeme von gewöhnlichen Differentialgleichungen und Anwendung auf das REENTRY-Problem. Numerische Math. 24, 249-275.

4. C.G. Simader (1972): On Dirichlet's boundary value problem. Lecture Notes 268. Springer-Verlag. Berlin-Heidelberg-New York.

5. J. Schröder (1975): Upper and lower bounds for solutions of generalized two-point boundary value problems Numer. Math. 23, 433-457.

6. J. Schröder (1977): Inclusion statements for operator equations by a continuity principle. Manuscripta Math. 21, 135-171.

7. J. Schröder (1977): Pointwise norm bounds for systems of ordinary differential equations. Report 77-14, Mathematisches Institut, Universität Köln.

*Differential Equations and Applications*
*W. Eckhaus and E.M. de Jager (eds.)*


# NONLINEAR BOUNDARY VALUE PROBLEMS ARISING IN PHYSICS

Roger TEMAM

Laboratoire d'Analyse Numérique et Fonctionnelle

C.N.R.S. et Université de Paris-Sud, 91405 - Orsay, France

---

The purpose of this lecture is to survey some recent results concerning a free boundary value problem which describes the equilibrium of a plasma confined in a Tokomak machine.

Section 1 contains the derivation of the equations and the general formulation of the problem. Section 2 gives the variational formulation of the problem and a statement of the main results for a particular problem which is however significant. Section 3 contains some indications on the results obtained in other cases, and finally Section 4 is devoted to open problems : among these let us already mention

- a singular perturbation problem
- two problems of bifurcation and existence of multiple solutions in non standard situations : one of these is related to the case where the perturbation is not smooth (and in some instances may not be small).

## 1. The governing equations.

The Tokomak machine is an axisymmetric one which is made schematilly of an axisymmetric shell which contains in its interior an annulus (a torus) of plasma.

Let Oz denotes the axis ; in a cross section plane Oxz we denote by $\Omega$ the region limited by the cross section $\Gamma$ of the shell ; $\Gamma$ is the boundary of $\Omega$. We denote by $\Omega_p$ the cross section of the plasma and $\Gamma_p$ the boundary of the plasma ; the region $\Omega_v = \Omega \setminus (\Omega_p \cup \Gamma_p)$ is empty.

In order to sligthly simplify the problem, we will consider instead of this toroïdal machine an infinite cylindar of axis Oz and cross-section $\Omega$ ; $\Omega_p$, $\Gamma_p$, $\Omega_v$, $\Gamma$ has the same signification as before. The reader is referred to [17] for the toroïdal case and for more details concerning the derivation of the equations.

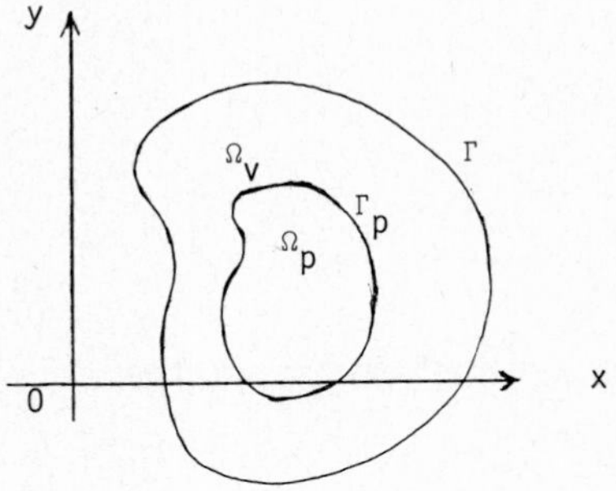


Equations

In the vacuum, we have the Maxwell equations

(1.1) $$\text{div } \vec{B} = 0 \text{ in } \Omega_v$$

(1.2) $$\text{curl } \vec{B} = 0 \text{ in } \Omega_v$$

where $\vec{B} = B_x \vec{i} + B_y \vec{j}$ ; (1.1) means

$$\frac{\partial B_x}{\partial x} + \frac{\partial B_y}{\partial y} = 0$$

so that there exists a locally defined function u (the flux function) such that

$$B_x = \frac{\partial u}{\partial y} , \qquad B_y = - \frac{\partial u}{\partial x} .$$

It will follow from the boundary condition that $u$ is a single valued function in $\Omega_v$ and (1.2) gives then

(1.3) $$\Delta u = 0 \text{ in } \Omega_v .$$

In the plasma, the governing equations are the magneto hydrodynamic equations (M.H.D.). In the absence of motion; the M.H.D. equations reduce to :

(1.4) $$\text{div } B = 0 \text{ in } \Omega_p ,$$

(1.5) $$\operatorname{curl} B = \mu_0 J \quad \text{in} \quad \Omega_p ,$$

(1.6) $$\nabla p = J \times B \quad \text{in} \quad \Omega_p ,$$

where $B, J, p$ are the magnetic flux, the current and the pressure, $B = B_x \vec{i} + B_y \vec{j}$, $J = J_x \vec{i} + J_y \vec{j}$ , $B, J, p$ depending only on $x$ and $y$ . We infer from (1.4) the existence of a flux function $u$ defined in $\Omega_p$ , with

$$B_x = \frac{\partial u}{\partial y} , \quad B_y = - \frac{\partial u}{\partial x} \quad \text{in} \quad \Omega_p .$$

The boundary conditions will show that $u$ is single-valued in $\Omega_p$ too. Then after an easy calculus, it follows from (1.5) and (1.6) that

$$\nabla p \parallel \nabla u , \text{ i.e. } p \text{ depends only on } u , \; p = g(u) ,$$

and

(1.7) $$\Delta u = - \mu_0 g'(u) \quad \text{in} \quad \Omega_p .$$

The function $g$ which plays the role of a <u>constitutive function</u> for the plasma must be considered as given. This function depends on the history of the plasma, i.e. the transient period before the plasma attains its equilibrium.

In Section 2 we will consider the simplest expression for $g$ : a quadratic function $g(u) = a_0 + a_1 u + a_2 u^2$ , where $a_0$ can be easily assumed to be equal to zero ( $p$ defined up to the addition of a constant), and where $a_1 = 0$ because of a physical constraint. Hence, setting $\lambda = 2\mu_0 a_2 > 0$ , (1.7) becomes

(1.8) $$\Delta u + \lambda u = 0 \quad \text{in} \quad \Omega_p .$$

<u>Boundary conditions</u>.

Let $\nu$ denote the unit outward normal on $\Gamma_p$ or $\Gamma$ and $\tau$ denote the unit tangent. Then $B.\nu = 0$ on $\Gamma_p$ and $\Gamma$ and $B.\tau$ is continuous across $\Gamma_p$ . We obtain that $\frac{\partial u}{\partial \tau} = 0$ on $\Gamma_p$ and $\Gamma$ . Since $u$ is defined up to the addition of a constant, we take $u = 0$ on $\Gamma_p$ and then $u$ = constant (an unknown constant) on $\Gamma$. We also obtain that $\frac{\partial u}{\partial \nu}$ is continuous across $\Gamma_p$ .

(1.9) $$\begin{cases} u = 0 \text{ on } \Gamma_p , \quad \frac{\partial u}{\partial \nu} \text{ continuous across } \Gamma_p \\ u = \text{(unknown) constant on } \Gamma . \end{cases}$$

Other physical conditions.

We have also some physical assumptions (cf. C. Mercier [10]) which impose

$$(1.10) \qquad u \neq 0 \text{ in } \Omega_p$$

$$(1.11) \qquad \int_{\Gamma_p} \frac{\partial u}{\partial \nu}\, d\ell = I \qquad \left( = \int_{\Gamma} \frac{\partial u}{\partial \nu}\, d\ell \right)$$

where $I > 0$ is a given number (the total current in the plasma).

## 2. The simple model.

Let $\Omega$ be an open bounded set of $R^n$ ($n = 2$ is the relevant case). We consider the free boundary value problem

$$(2.1) \qquad \Delta u = 0 \text{ in } \Omega_v$$

$$(2.2) \qquad \Delta u + \lambda u = 0 \text{ in } \Omega_p$$

$$(2.3) \qquad u = 0 \text{ on } \Gamma_p$$

$$(2.4) \qquad \frac{\partial u}{\partial \nu} \text{ continuous across } \Gamma_p$$

$$(2.5) \qquad u = \text{(unknown) constant on } \Gamma$$

$$(2.6) \qquad \int_{\Gamma} \frac{\partial u}{\partial \nu}\, d\ell = I$$

$$(2.7) \qquad u \neq 0 \text{ in } \Omega_p ,$$

where $I > 0$ , $\lambda > 0$ are given, the function $u$ and the open set $\Omega_p$ are unknown.

Variational formulation.

We first observe that, because of the maximum principle, $u$ has in $\Omega_v$ the sign of $u_{|\Gamma}$ , and by the strong maximum principle and (2.6) ($I > 0$) this sign must be the positive one. Then because of (2.7)-(2.4) and the strong maximum principle, $u < 0$ in $\Omega_p$ , and hence we have a simple caracterisation of $\Omega_p$ , $\Gamma_p$ and $\Omega_v$ in term of $u$

$$(2.8) \qquad \begin{cases} \Omega_v = \{x \in \Omega,\ u(x) > 0\} \\ \Omega_p = \{x \in \Omega,\ u(x) < 0\} \\ \Gamma_p = \{x \in \Omega,\ u(x) = 0\} \end{cases}$$

We consider now the Sobolev space $H^1(\Omega)$ and its subspace

$$W = \{v \in H^1(\Omega) \text{ , } v_{|\Gamma} = \text{constant}\}$$

which is a closed one, and $W$ is isomorphic to $H_0^1(\Omega) \oplus \mathbf{R}$ . For $v \in W$ , the value of $v$ on $\Gamma$ is denoted $v(\Gamma)$ .

Clearly $u \in W$ . Now if $v$ is a test function in $W$ , we multiply (2.1) and (2.2) by $v$ , integrate over $\Omega_v$ and $\Omega_p$ respectively and add the equalities which we obtain. Using (2.4)-(2.5)-(2.6) and the Green formula, we easily obtain :

$$(\nabla u, \nabla v) = \lambda \int_{\Omega_p} u \ v \ dx = I \ v(\Gamma) \ .$$

But

$$\int_{\Omega_p} u \ v \ dx = - \int_{\Omega} (u_-) \ v \ dx \ ,$$

where $s_- = \max(-s,0)$ . Whence,

(2.9) $$(\nabla u, \nabla v) + \lambda(u_-, v) = I \ v(\sigma) \ ,$$

where $(.,.)$ denotes the scalar product in $L^2(\Omega)$ or $L^2(\Omega)^n$ and $\nabla u = \text{grad } u$ .

Finally if $u$ is a solution of (2.1)-(2.7) then $u \in W$ and (2.9) holds for every $v \in W$ . It is easy to see that conversely if $u \in W$ satisfies (2.9) for each $v \in W$ , if we define $\Omega_p$, $\Gamma_p$, $\Omega_v$ through (2.8) then (2.1)-(2.7) are satisfied provided $\Gamma_p$ is a sufficiently regular curve. We call Problem I the problem of finding $u$ (and $\Omega_p$) solution of (2.1)-(2.7) and Problem II the problem of finding $u \in W$ satisfying (2.9) for each $v \in W$ . Then Problem II is the weak or variational formulation of Problem I, and we will mainly consider Problem II.

Remark 2.1. In Problem I , $u_{|\Omega_v}$ appears as the first eigenfunction of the Dirichlet problem in $\Omega_v$ , and $\lambda$ the corresponding (first) eigenvalue. Since $\Omega_v \subset \Omega$ , we have

(2.10) $$\lambda > \lambda_1$$

where $\lambda_1$ is the first eigenvalue of the Dirichlet problem in $\Omega$ (cf. Courant-Hilbert [ 4 ]).

If $u$ is the solution of Problem II, the set $\Omega_v$ may be empty for an arbitrary $\lambda$ ; however $u(\Gamma) > 0$ and $\Omega_v$ is not empty if (2.10) holds.

From now one, we assume that (2.10) is satisfied.

It is easy to see that Problem II is equivalent to one of the following problems

Problem III

$$(2.11)\qquad \begin{cases} -\Delta u + \lambda u_- = 0 \text{ in } \Omega \\ u = \text{(unknown) constant on } \Gamma \\ \int_\Gamma \frac{\partial u}{\partial \nu} d\ell = I \end{cases}$$

Problem IV

To find $u$ , critical value in $W$ of the functional

$$(2.12)\qquad e(v) = \frac{1}{2} |\nabla v|^2 - \frac{\lambda}{2} |v_-|^2 - Iv(\Gamma)$$

where $|.|$ is the norm in $L^2(\Omega)$ or $L^2(\Omega)^n$ .

Note that the functional $e$ is unbounded in $W$ :
$e(v) \to +\infty$ for $v = \xi v_0$ , $v_0 > 0$ and $\xi \to +\infty$ ,
$e(v) \to -\infty$ for $v \equiv \xi$ , $\xi \in R$, $\xi \to -\infty$ .
Therefore a critical point cannot be a minimum nor a maximum (see below).

Existence and regularity results.

Théorème 2.1. For $I > 0$ and $\lambda > \lambda_1$ given, Problem II possesses at least one solution $u \in W$ . The function $u$ belongs to the Sobolev space $W^{3,\alpha}(\Omega)$ for all $\alpha \geqslant 1$ and to $\mathscr{C}^{2,\eta}(\bar{\Omega})$ for all $\eta$ satisfying $0 \leqslant \eta < 1$ .

Let the sets $\Omega_p$, $\Omega_v$, $\Omega_p$, be defined by (2.8). Then $u$ satisfies (2.1)-(2.2)-(2.3)-(2.5)-(2.6)-(2.7) and $u$ is analytic in $\Omega_p$ and $\Omega_v$ . In the neighborhood of each point $x \in \Gamma_p$ such grad $u(x) \neq 0$ , $\Gamma_p$ is a $\mathscr{C}^2$ curve and (2.4) is satisfied.

The proof is given in [18]. For the existence we solve Problem IV. We observe that a possible solution $u$ satisfies

$$\lambda \int_\Omega u_- \, dx = \int_r \Delta u \, dx = \int_\Gamma \frac{\partial u}{\partial \nu} d\ell = I$$

and therefore belongs to the set

$$K = \{u \in W , \int_\Omega u_- \, dx = \frac{I}{\lambda} \}.$$

The idea of the proof of existence is to show that $e$ is bounded from below on $K$ and attains its minimum at a point $u$. Then $u$ appears as the solution of a constrained minimisation problem for which the Lagrange multiplier happens to be $0$ so that

$$e'(u) = 0,$$

and $u$ is a solution to Problem IV.

The regularity results follows from consideration of equation (2.11) and classical regularity results for the Dirichlet problem.

■

Concerning the global regularity of $\Gamma_p$ we have a precise result which follows directly from a theorem of Hartman and Wintner [8] when $n = 2$.

<u>Theorem 2.2.</u> For $n = 2$, $\Gamma_p$ is a piecewise $\mathscr{C}^2$ curve, with at most a finite number of Lipschitz discontinuities.

<u>Remark 2.2.</u> In the case of the solution $u$ obtained by minimising $e$ on $K$, $\Omega_p$ is a connected open set. If $n = 2$, $\Gamma_p$, for this solution is globally a $\mathscr{C}^2$ curve.

In all cases, we have a more precise local regularity result for $\Gamma_p$ due to D. Kinderlehrer and L. Nirenberg [9].

<u>Theorem 2.3.</u> In the neighbouhood of every point $x \in \Gamma_p$ where $\text{grad}\, u(x) \neq 0$, $\Gamma_p$ is an analytic curve and $u$ is analytic in the intersection of this neighbouhood with $\overline{\Omega}_p$ and in the intersection of this neighbouhood with $\overline{\Omega}_v$. [1]

<u>Remark 2.3.</u> An alternate proof of existence of solution of Problem II has been given by Berestycki and Brezis [3] using a variational principle involving higher derivatives and different from ours. Damlamian [5] has shown that the variational principle used in [3] is dual to the variational problem considered here, in the sense of the non convex duality theory of J. Toland [19].

<u>Uniqueness and Non Uniqueness results.</u>

Let $\lambda_i = \lambda_i(\Omega)$, $i \geq 1$, denote the sequence of eigenvalue of the Dirichlet problem in $\Omega$. Then

---

[1] Note that $u$ <u>is not</u> $\mathscr{C}^\infty$ across $\Gamma_p$.

Theorem 2.4. For every $I > 0$ and for $\lambda < \lambda_2$, the solution of Problem II is unique. The solution is unique for $\lambda \leq \lambda_2$ if $\lambda_1 < \lambda_2$.

The proof is given in [18] when the dimension of space is $n = 2$. The proof has been simplified and extended to higher dimensions by Puel [14] who gives also in [14] an alternate proof of existence based on a result of Ambrosetti and Rabinowitz [1].

A non uniqueness result of D. Schaeffer [15] shows that this uniqueness result is "almost" optimal.

Theorem 2.5. For $n = 2$, there exists $\Omega \subset \mathbb{R}^2$ and $\lambda$ such that

$$\lambda_2(\Omega) < \lambda < \lambda_3(\Omega), \tag{2.13}$$

and for these $\Omega$ and $\lambda$, Problem II possesses at least two solutions.

The proof in [15] consists in the construction of such a pair $\lambda, \Omega$ ; $\Omega$ has the shape indicated in the figure below with $\varepsilon$ sufficiently small

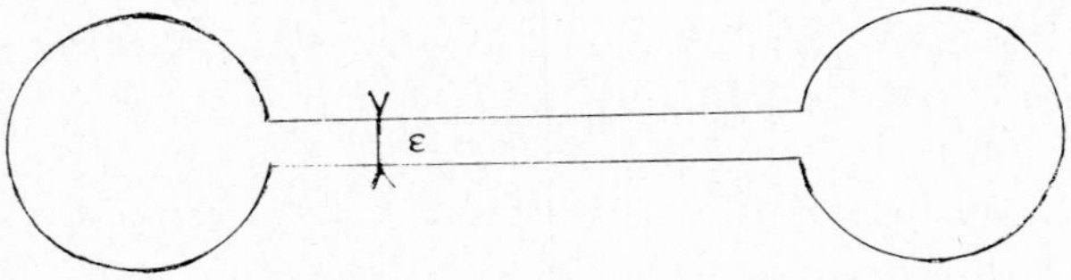


Remark 2.4. The proof of Theorem 2.5 given in [18] also shows that if non uniqueness occurs in the neighbourhood of a pair of solution $u_*$, $\lambda_*$, then there exists $v \not\equiv 0$ such that

$$\begin{cases} \Delta v + \lambda_* h(-u_*)\, v = 0 \quad \text{in } \Omega \\ v = 0 \quad \text{on } \Gamma \\ \displaystyle\int_\Gamma \frac{\partial v}{\partial \nu}\, d\ell = 0 \end{cases} \tag{2.14}$$

($h$ the Heaviside function, arbitrarily defined at 0) and this means in particular that $\lambda_*$, $v_*$ are solution of the two following eigenvalue problems :

(2.15)
$$\begin{cases} \Delta v + \mu h(-u_*)\, v = 0 & \text{in } \Omega \\ v = 0 & \text{on } \Gamma \end{cases}$$

(2.16)
$$\begin{cases} \Delta v + \mu h(-u\ )\, v = 0 & \text{in } \Omega \\ v = \text{constant on } \Gamma \\ \int_\Gamma \frac{\partial v}{\partial \nu}\, d\ell = 0 . \end{cases}$$

## 3. Other Models

The model studied in Section 2 corresponds to a simple choice of the function $g$, $p = g(u)$. We now discuss some other type of functional dependance of $p$ in term of $u$.

### 3.1. Arbitrary g's.

Let $g$ be a regular ($\mathscr{C}^2$) function from $\mathbb{R}$ into $\mathbb{R}$ such that

(3.1)
$$g'(0) = 0$$

and

(3.2)
$$g'(u) > 0 \quad \text{for} \quad u < 0 .$$

We may consider the case where $p = g(u)$ ; the problem is then similar to Problem I, (2.2) being replaced by (1.7). With suitable assumptions on the growth of $g$ at infinity (cf. [17]) we obtain an existence result as in Theorem 2.1. The regularity results are essentially the same as in Theorem 2.1 and 2.3, provided $g$ is sufficiently regular. Finally an uniqueness result similar to Theorem 2.5 holds for $\lambda$ sufficiently small.

### 3.2. Problem with λ fixed.

An interesting variant of Problem I corresponds to the case where $\lambda$ is not specified but instead, as in a linear eigenvalue problem, the value of the functional

$$\int_{\Omega_p} u^2\, dx = \int_\Omega (u_-)^2\, dx$$

is specified $(= \kappa > 0)$ .

In [17], this problem is reduceed to the problem of finding a critical value of

(3.3) $$e_1(v) = \frac{1}{2}\,|\nabla v|^2 - Iv(\Gamma)$$

on the subset of W ,

(3.4) $$\Sigma(\kappa) = \{v \in W\ ,\ \int_\Omega (u_-)^2\,dx = \kappa\}$$

The function $e_1$ is bounded from below on any set $\Sigma(\kappa)$ and attains its minimum at a point u which is solution of the problem. The regularity of the solution is obtained as in Section 2. Concerning uniqueness, C. Guillopé [7] has shown that the critical point is unique when

(3.5) $$\frac{I}{\kappa} \text{ is sufficiently small.}$$

We do not know if multiple solutions may appear when $\frac{I}{\kappa}$ is large.

3.3. The Grad-Mercier Models.

In these models, the functional dependance of p in term of u is more complicated. For a given real function u defined on $\Omega$ , we consider the function

$$\overline{\beta}(u)(x) = \text{mes}\{y, u(y) \leqslant u(x)\}\ .$$

The operator $\overline{\beta}$ is not monotone, nor local. The function

$$p(u) = g(\overline{\beta}(u))$$

is one of the functional dependance of p in term of u arising in the Grad-Mercier models [6] , [11].

Problems of this type has been systematically studied by J. Mossino [12] in her thesis ; cf. also [20] and [13]. Existence, approximation and regularity results are obtained using the following tools :

- quasi-variational inequalities [2]
- monotone sequences of sub or super solutions
- convex analysis and optimisation.

The reader is refered to [12], [13] , [20] for the details.

## 4. Open Problems

4.1. Is the solution of Problem I.IV unique when $\Omega$ is convex ? (Conjecture of D. Schaeffer, true for simple cases).

4.2. Can one show in Problem I-IV the apparition of a bifurcating branch of solutions near some $\lambda > \lambda_*$ ? (The classical results do not apply since $f(u) = u_-$ is not smooth).

4.3. Numerical computations made by M. Sermange [16] seem to indicate that there is a vertical branch of solutions at a point $\lambda_* \in ]\lambda_2, \lambda_3[$ (notations of Theorems 2.3, 2.5, $\Omega$ has the same shape as for Theorem 2.5)

Is there a vertical bifurcation ?

4.4. Study the behaviour of the solutions $u = u(\lambda)$ of Problems I-IV when $\lambda = \varepsilon^{-1} \longrightarrow +\infty$ . This is a non stand singular perturbation problem - ($\lambda \longrightarrow +\infty$ means that the pressure becomes infinite and it is expected that the plasma shrinks to one point : this is easy to see, using Bessel functions in the one dimensional model).

4.5. The study of the existence of multiple solutions for the model considered in Section 3.2 is related to the problem of bifurcation for the equation of critical points of a functional on a manifold.

## References

[1] Ambrosetti , P. Rabinowitz -
J. Funct. Anal., 14, 1973, p.349-381.

[2] Bensoussan A., J.L. Lions -
Comptes Rendus, 276, série A, 1973, p.1189-1192.

[3] Berestycki H., H. Brezis -
Sur certains problèmes de frontière libre, C.R. Ac. Sc., 283, Série A, 1976, p.1091-1094.

[4] Courant R., D. Hilbert -
Methods of Mathematical Physics, Interscience Publishers, New York, 1953.

[5] Damlamian A. -
Exposé au séminaire d'Equations aux dérivées partielles non linéaires, Orsay, 1977.

[6] Grad H., P.N. Hu, D.C. Stevens -
Adiabatic evolution of plasma equilibrium, Proc. Nat. Ac. Sc. U.S.A., vol.72, N°10, 1975, p.3789-3793.

[7] Guillopé C. -
Thèse de 3ème cycle, Université de Paris-Sud, Orsay, 1977.

[8] Hartman P., A. Wintner -
On the local behavior of solutions of non parabolic partial differential equations, Amer. J. Math., 85, 1953, p.449-476.

[9] Kinderlehrer D., L. Nirenberg -
Regularity in free Boundary Problems, Annali Scuola Norm. Sup. Pisa, 25, 1977, p.373-391.

[10] Mercier C. -
The magnetohydrodynamic approach to the problem of plasma confinment in closed magnetic configuration, Publication EURATOM-CEA, Luxembourg, 1974.

[11] Mercier C. -
Communication personnelle.

[12] Mossino J. -
Etude de quelques problèmes non linéaires d'un type nouveau apparaissant en physique des plasmas, Thèse, Université de Paris Sud, Orsay, 1977.
Cf. also Comptes Rendus Ac. Sc. 282, S rie A 1976, p.187 and an article to appear in the Israel Journ. of Math.

[13] Mossino J., J.P. Zolésio -
Comptes Rendus Acad. Sc., Série A, 1977.

[14] Puel J.P. -
Sur un problème de valeur propre non linéaire et de frontière libre, C.R. Ac. Sc., 284, Série A, 1977, p.861-863.

[15] Schaeffer D. -
Non-uniqueness in the Equilibrium Shape of a Confined Plasma, Comm. Part. Diff. Equ., vol.2, n°6, 1977, p.587-600.

[16] Sermange M. -
Rapport IRIA, to appear.

[17] Temam R. -
A nonlinear eigenvalue problem : the shape at equilibrium of a confined plasma, Arch. Rat. Mech. Annal., 60, 1976, p.51-73.

[18] Temam R. -
Remarks on a free boundary value problem arising in plasma physics, Comm. in Part. Diff. Equ., vol.2, n°6, 1977, p.563-586.

[19] Toland J. -
A duality principle for nonconvex optimisation and the Calculus of variations, Arch. Rat. Mech. Anal., To appear.

[20] Mossino J., R. Temam -
Certains problèmes non linéaires de la physique des plasmas, in Mathematical Aspects of Finite Element Methods, Proceedings of a Conference held at Rome, I. Galligani and E. Magenés editors, Lecture Notes in Mathematics, vol. 606, Springer Verlag 1977.

*Differential Equations and Applications*
*W. Eckhaus and E.M. de Jager (eds.)*
*©North-Holland Publishing Company (1978)*

# ON THE EXISTENCE OF CLASSICAL SOLUTIONS TO AN ELLIPTIC FREE BOUNDARY PROBLEM

Bernard A. Fleishman†
Department of Mathematical Sciences
Rensselaer Polytechnic Institute
Troy, New York 12181, U.S.A.

Thomas J. Mahar*
Department of Mathematics
Utah State University
Logan, Utah 84322, U.S.A.

An existence theorem is proved for a two-dimensional free boundary problem for a nonlinear Poisson equation. The method is constructive, employing a monotone iterative scheme, and it provides classical solutions.

## 1. INTRODUCTION

In this paper we prove an existence theorem for a two-dimensional free boundary problem for a nonlinear Poisson equation. Our method is constructive, employing a monotone iteration scheme, and it provides classical solutions.

Specifically, in the unit disc $D = \{(r,\theta): 0 \leq r < 1,\ 0 \leq \theta \leq 2\pi\}$ we consider the following problem, denoted by $\mathcal{Q}(\varepsilon)$:

$$\Delta u + f(u) = 0 \quad \text{in } D/\Gamma, \tag{1.1a}$$

$$u(1,\theta) = \varepsilon h(\theta), \quad 0 \leq \theta \leq 2\pi. \tag{1.1b}$$

where $\Gamma$ is a free boundary given by $r = \gamma(\theta)$. The function h is continuous and $2\pi$-periodic, and satisfies $0 < h(\theta) \leq 1$, while the function f is given by

$$f(u) = \begin{cases} 0, & u < \mu \\ 1, & \mu \leq u \end{cases} \tag{1.2}$$

† Research supported by the U.S. Army Research Office.

*Some of this work was done at the Courant Institute of Mathematical Sciences, New York University.

where $\mu$ is a positive real number. The set of points $(r,\theta)$ satisfying $u(r,\theta) = \mu$ constitutes the free boundary, which must be determined. A solution consists of a free boundary $\Gamma$ and a function $u(r,\theta)$ belonging to $C^1(\bar{D}) \cap C^2(D/\Gamma)$ which satisfies (1.1).

The iteration scheme we use is

$$\Delta u_{n+1} + f(u_n) = 0 \qquad \text{in } D/\Gamma_n, \qquad n = 0,1,\dots, \tag{1.3a}$$

$$u_{n+1}(1,\theta) = \varepsilon h_{n+1}(\theta), \quad 0 \le \theta \le 2\pi . \tag{1.3b}$$

This is a linear Poisson interface problem, the interface given by the curve $\Gamma_n = \{(r,\theta): u_n(r,\theta) = \mu\}$. The boundary functions $h_{n+1}(\theta)$, specified later, are such that $h_n \to h$ uniformly as $n \to \infty$.

Problems of the form (1.1) occur in the study of steady-state heat conduction. If $v$ denotes the temperature in $D$, $k(v)$ the conductivity, and $g(v)$ a source term, then $v$ satisfies the equation

$$\operatorname{div}(k(v)\operatorname{grad} v) + g(v) = 0. \tag{1.4}$$

If there is a phase change involved, $k$ and $g$ may have jump discontinuities. The Kirchhoff transformation $u = \int_0^v k(s)ds$ then leads to an equation of the form (1.1) for $u$. We have taken the simplest possible function $f$ which has a jump discontinuity.

Problems of the form (1.1) also arise from attempts to approximate solutions to similar problems with smooth nonlinearities. If $\hat{f}(u)$ is almost constant on either side of $\mu$, and has a rapid change in a small neighborhood of $\mu$, the function $\hat{f}(u)$ can be approximated by one having the form (1.2). The resulting problem may be easier to solve, and thus can be used to determine approximations to solutions of the original problem.

In Section 2 we study two symmetric problems related to $Q(\varepsilon)$. Setting $\varepsilon = 0$, we obtain $Q(0)$, given by

$$\Delta u + f(u) = 0 \qquad \text{in } D/\Gamma \tag{1.5a}$$

$$u(1,\theta) \equiv 0, \tag{1.5b}$$

whereas setting $h(\theta) \equiv 1$ gives the problem

$$\Delta u + f(u) = 0 \quad \text{in } D/\Gamma \tag{1.6a}$$

$$u(1,\theta) \equiv \varepsilon. \tag{1.6b}$$

Both of these problems reduce to ordinary differential equations for solutions symmetric about the origin. A non-trivial solution of (1.5), $\bar{u}_o$, is used as the first member of a monotone sequence of iterates, $\{u_n(r,\theta)\}$, bounded above by a solution of (1.6), $\bar{u}_\varepsilon$. The monotonicity and boundedness of this sequence, as well as of the associated sequence of interfaces $\{\Gamma_n\}$, is insured by an interplay of a minimum principle and the fact (which is not obvious but is shown) that, for small enough $\varepsilon$, in a certain annulus in D, $r(0) \leq r \leq r(\varepsilon)$, every $u_n$ is strictly decreasing with r (for fixed $\theta$).

In Section 3 we prove our minimum principle (Theorem 1) and a uniqueness result (Theorem 2) for linear Poisson interface problems. Properties of the iterates $u_n$ and $\Gamma_n$, established in Section 4, are used in the convergence proof in Section 5, which yields our main result (Theorem 3), on the existence of solution to the free boundary problem $\mathcal{Q}(\varepsilon)$.

In recent work [1] on the role of order structures in nonlinear functional analysis, Amann notes that in this way broader results can be obtained for discontinuous nonlinear boundary value problems than the authors and Chandra have achieved (in [2, 3, 4]) by relying on monotone iteration schemes (as is done in this paper). The latter technique, however, yields existence proofs which are constructive (in the usual sense), and apparently this is not the case in [1].

Further remarks on our results are contained in Section 6. Finally, a few results used in the analysis (Lemmas A1-2-3) are collected in an appendix.

## 2. SYMMETRIC PROBLEMS

Because the boundary conditions are independent of $\theta$, we may seek solutions of the form $u = u(r)$ to problems (1.5) and (1.6). Then (1.5) reduces to

$$(ru')' + rf(u) = 0, \qquad 0 < r < 1, \tag{2.1a}$$

$$u'(0) = u(1) = 0 . \tag{2.1b}$$

By an analysis similar to that in [2,3,4], it can be shown that any solution $u(r)$ for (2.1) satisfies the integral equation

$$u(r) = \int_r^1 \frac{1}{t} \int_0^t s\, f(u(s))ds\, dt. \tag{2.2}$$

Differentiating (2.2), we find that

$$u'(r) = -\frac{1}{r} \int_0^r s\, f(u(s))ds. \tag{2.3}$$

Since $u' \leq 0$, $u$ has a maximum at $r = 0$; if $u(0) > \mu$, then $u'(r) < 0$ for $0 < r \leq 1$, in which case there is at most one value of $r$ for which $u = \mu$, this value of $r$ defining a free boundary for (2.1).

A similar procedure applied to the problem (1.6) leads to

$$(ru')' + rf(u) = 0, \qquad 0 < r < 1, \tag{2.4a}$$

$$u'(0) = 0, \quad u(1) = \varepsilon . \tag{2.4b}$$

Any solution to (2.4) satisfies the relations

$$u(r) = \varepsilon + \int_r^1 \frac{1}{t} \int_0^t s\, f(u(s))ds\, dt, \tag{2.5}$$

$$u'(r) = -\frac{1}{r} \int_0^r s\, f(u(s))ds \leq 0. \tag{2.6}$$

Again, $u$ has a maximum at $r = 0$, and $u(0) > \mu$ implies that $u' < 0$. Thus, any solution to (2.4) has at most one free boundary.

Let $u_o(r)$ denote a solution to (2.1), and $u_\varepsilon(r)$ a solution to (2.4), with respective free boundaries $r(0)$ and $r(\varepsilon)$ (when they exist). Since $f(u)$ given in (1.2) is so simple, we can give explicit representations for these solutions. The solutions to (2.1) are:

i) $u_o(r) \equiv 0, \qquad 0 \leq r \leq 1$

ii) $$u_o(r) = \begin{cases} -\dfrac{r^2(0)}{2} \ln r(0) + \dfrac{r^2(0)}{4} - \dfrac{r^2}{4}, & 0 \leq r \leq r(0) \\ -\dfrac{r^2(0)}{2} \ln r, & r(0) \leq r \leq 1. \end{cases}$$

The free boundary $r(0)$ in $(0,1)$ is found by solving the transcendental equation $u_o(r) = \mu$. Setting $r^2 = \xi$, we write this equation as

$$- \xi \ln \xi = 4\mu, \qquad 0 < \xi < 1. \tag{2.7}$$

Since $0 \leq - \xi \ln \xi \leq e^{-1}$ for $0 \leq \xi \leq 1$, there are two possible choices for the free boundary $r(0)$ when $4\mu < e^{-1}$, there is a unique $r(0)$ when $4\mu = e^{-1}$, and there is none when $4\mu > e^{-1}$.

We shall assume $4\mu < e^{-1}$. Using the two possible values of $r(0)$ in ii), we obtain two distinct solutions. Let $\underline{u}_o(r)$ denote the zero solution, $\bar{u}_o(r)$ the solution from ii) corresponding to the larger $r(0)$, and $u_o(r)$ the remaining solution. By techniques similar to those used in [2,3,4] it can be shown that

$$\underline{u}_o(r) \leq u_o(r) \leq \bar{u}_o(r), \qquad 0 \leq r \leq 1. \tag{2.8}$$

The solutions to (2.4) are as follows:

iii) $u_\varepsilon(r) \equiv \varepsilon , \qquad 0 \leq r \leq 1,$

$$\text{iv)} \quad u_\varepsilon(r) = \begin{cases} - \dfrac{r^2(\varepsilon)}{2} \ln r(\varepsilon) + \dfrac{r^2(\varepsilon)}{4} + \varepsilon - \dfrac{r^2}{4}, & 0 \leq r \leq r(\varepsilon), \\ - \dfrac{r^2(\varepsilon)}{2} \ln r + \varepsilon , & r(\varepsilon) \leq r \leq 1 . \end{cases}$$

Setting $r^2(\varepsilon) = \xi$ , we find $r(\varepsilon)$ in $(0,1)$ by solving

$$- \xi \ln \xi = 4(\mu - \varepsilon) . \tag{2.7'}$$

Since we are assuming $4\mu < e^{-1}$, (2.7') also has two roots, and so there are two possible values for $r(\varepsilon)$ in iv). Let $\underline{u}_\varepsilon(r) \equiv \varepsilon$, and let $u_\varepsilon(r)$ and $\bar{u}_\varepsilon(r)$ denote the solutions from iv) corresponding to the smaller and larger root, respectively, of (2.7'). Then

$$\underline{u}_\varepsilon(r) \leq u_\varepsilon(r) \leq \bar{u}_\varepsilon(r), \qquad 0 \leq r \leq 1. \tag{2.9}$$

Techniques similar to those used in [2,3,4] can be used to show that

$$\underline{u}_o(r) \le \underline{u}_\varepsilon(r) , \qquad 0 \le r \le 1 , \tag{2.10a}$$

$$\bar{u}_o(r) \le \bar{u}_\varepsilon(r) , \qquad 0 \le r \le 1 . \tag{2.10b}$$

Inequality (2.10a) is trivial here, but we can improve on (2.10b), as follows. Since $f(u)$ is nondecreasing in $u$, (2.10b) implies $f(\bar{u}_o(r)) \le f(\bar{u}_\varepsilon(r))$; using this in (2.3) and (2.6), we find $\bar{u}'_\varepsilon(r) \le \bar{u}'_o(r)$. Integrating between $r$ and 1, we obtain

$$\bar{u}_o(r) + \varepsilon \le \bar{u}_\varepsilon(r). \tag{2.11}$$

Using (2.11) and results analogous to those in [4], we find

$$r(\varepsilon) > r(0), \qquad \varepsilon > 0 , \tag{2.12a}$$

$$r(\varepsilon) \to r(0) \qquad \text{as } \varepsilon \to 0. \tag{2.12b}$$

These results can also be derived by studying (2.7) and (2.7'). Indeed, since $4\mu < e^{-1}$, applying the implicit function theorem to (2.7') shows that $r(\varepsilon)$ is a differentiable function of $\varepsilon$ for $0 \le \varepsilon \le \varepsilon_o$, where $\varepsilon_o$ is any positive number less than $\mu$. Consequently (2.12b) can be improved to read as follows:

$$0 \le r(\varepsilon) - r(0) \le k_1\varepsilon , \qquad 0 \le \varepsilon \le \varepsilon_o, \tag{2.13}$$

where $k_1$ is a positive real number.

We make a final observation about the solutions $\bar{u}_o$ and $\bar{u}_\varepsilon$. By use of the explicit representations ii) and iv) along with (2.13), it can be shown that

$$0 \le \bar{u}_\varepsilon(r) - \bar{u}_o(r) \le k_2\varepsilon , \qquad 0 \le r \le 1, \tag{2.14}$$

where $k_2$ is a positive real number independent of $r$ and $\varepsilon$. Estimate (2.14) is established by considering the representations for $\bar{u}_o$ and $\bar{u}_\varepsilon$ on each of the subintervals $0 \le r \le r(0)$, $r(0) \le r \le r(\varepsilon)$, and $r(\varepsilon) \le r \le 1$.

## 3. MINIMUM PRINCIPLE AND UNIQUENESS THEOREM

In order to prove that our iteration scheme is monotone, we derive a minimum principle for linear Poisson interface problems. If $u_{n-1} \leq u_n$ in $\bar{D}$, (1.3) implies $\Delta(u_{n+1} - u_n) \leq 0$ at all points in D not on $\Gamma_{n-1}$ or $\Gamma_n$. We now develop a minimum principle which guarantees $u_{n+1} - u_n \geq \varepsilon \min (h_{n+1} - h_n)$ throughout $\bar{D}$.

Theorem 1. Let $\Gamma = \Gamma_1 \cup \Gamma_2$, where $\Gamma_1 = \{(r,\theta): r = \gamma_1(\theta)\}$, $\Gamma_2 = \{(r,\theta): r = \gamma_2(\theta)\}$ , and the function $\gamma_i$ are $C^2$ and $2\pi$-periodic and satisfy $0 < \gamma_1(\theta) < \gamma_2(\theta) < 1$ for $0 \leq \theta \leq 2\pi$. Let $w \in C^1(\bar{D}) \cap C^2(D/\Gamma)$ be such that

$$\Delta w \leq 0 \qquad \text{in } D/\Gamma , \tag{3.1a}$$

$$w(1,\theta) = G(\theta), \qquad 0 \leq \theta \leq 2\pi , \tag{3.1b}$$

with G continuous and $2\pi$-periodic. If $\alpha = \min G$, then $w \geq \alpha$ in $\bar{D}$.

Proof: Our argument uses the standard minimum principle (as developed in [7], for example). Suppose that $(\bar{r},\bar{\theta}) \in D$ is such that w has a local minimum there. We shall show that $w(\bar{r},\bar{\theta}) \geq \alpha$.

Let $D_1 = \{(r,\theta): 0 \leq r < \gamma_1(\theta)\}$ , $D_2 = \{(r,\theta): \gamma_1(\theta) < r < \gamma_2(\theta)\}$, and $D_3 = \{(r,\theta): \gamma_2(\theta) < r < 1\}$.

i) $(\bar{r},\bar{\theta}) \in D_3$. As $w \in C^2(D_3)$, $w \equiv$ constant $= c$ in $\bar{D}_3$. If $G(\theta) \not\equiv c$, $w = G$ on $\partial D$ gives a contradiction and $(\bar{r},\bar{\theta})$ cannot be a local minimum. If $G(\theta) \equiv c$, then $c = \alpha$ and $w(\bar{r},\bar{\theta}) = \alpha$.

ii) $(\bar{r},\bar{\theta}) \in \Gamma_2$. Since w has a local minimum at $(\bar{r},\bar{\theta})$, its restrictions to $\bar{D}_2$ and $\bar{D}_3$ have a local minimum there too. But, $\partial_{-r} w > 0$ and $\partial_r w > 0$ cannot both hold at $(\bar{r},\bar{\theta})$. Thus, $w \equiv$ constant $=c$ in $\bar{D}_3$. As in i), either $(\bar{r},\bar{\theta})$ is not a local minimum for w, or $w(\bar{r},\bar{\theta}) \geq \alpha$.

iii) $(\bar{r},\bar{\theta}) \in D_2$. We have $w \in C^2(D_2)$, so that $w \equiv c$ in $\bar{D}_2$. Since $w \in C^2(D_3)$, its minimum in $\bar{D}_3$ must occur on $\Gamma_2$ or $\partial D$. If the minimum occurs on $\Gamma_2$, $w \equiv c$ in $\bar{D}_3$ because $\partial_r w \equiv 0$ on $\Gamma_2$ ($w \equiv c$ in $\bar{D}_2$), and so $c = \alpha$. But if the minimum in $\bar{D}_3$ occurs on $\partial D$, and not on $\Gamma_2$, then $c \geq \alpha$. Hence, $(\bar{r},\bar{\theta}) \in D_2$ implies $w(\bar{r},\bar{\theta}) \geq \alpha$.

iv) $(\bar{r},\bar{\theta}) \in \Gamma_1$. As in ii), $w \equiv c$ in $\bar{D}_1 \cup \bar{D}_2$. The same reasoning used in iii) now applies.

v) $(\bar{r},\bar{\theta}) \in D_1$. Since $w \in C^2(D_1)$, $w \equiv c$ in $\bar{D}_1$ and $\partial_r w \equiv 0$ on $\Gamma_1$. A combination of the arguments used in i) - iv) shows that the minimum of w in $\bar{D}_2 \cup \bar{D}_3$ must occur on $\partial D$, and not on $\Gamma_1$ or $\Gamma_2$. Thus, $c \geq \alpha$ and $w(\bar{r},\bar{\theta}) \geq \alpha$. This completes the proof.

We have shown that if w has a local minimum in D at $(\bar{r},\bar{\theta})$, then $w(\bar{r},\bar{\theta}) \geq \alpha = \min G = \min w$ on $\partial D$. If w achieves its global minimum in $\bar{D}$ at a point $(\hat{r},\hat{\theta})$, then $w(\hat{r},\hat{\theta}) \geq \alpha$ if $(\hat{r},\hat{\theta}) \in D$ (since a global minimum is also a local one) by the above, while $w(\hat{r},\hat{\theta}) \geq \alpha$ is true if $(\hat{r},\hat{\theta}) \in \partial D$ because $w = G$ on $\partial D$.

Similar reasoning leads to a maximum principle if $\Delta w \geq 0$ in $D/\Gamma$. A combination of both principles implies the following uniqueness theorem which covers linear Poisson interface problems such as (1.3).

Theorem 2. Suppose the assumptions in Theorem 1 hold, except that the inequality (3.1a) is replaced by the equality

$$\Delta w = 0 \text{ in } D/\Gamma. \tag{3.2}$$

Then w is the only solution in $C^1(\bar{D}) \cap C^2(D/\Gamma)$ of BVP (3.2), (3.1b).

## 4. PROPERTIES OF THE ITERATES

We now study the iteration procedure

$$\Delta u_{n+1} + f(u_n) = 0 \quad \text{in } D/\Gamma_n, \qquad n = 0,1,\ldots \tag{1.3a}$$

$$u_{n+1}(1,\theta) = \varepsilon h_{n+1}(\theta), \; 0 \leq \theta \leq 2\pi . \tag{1.3b}$$

With $\alpha \equiv \min h(\theta) > 0$, let us define the functions $h_n(\theta)$ by

$$h_o(\theta) \equiv 0 \tag{4.1a}$$

$$h_n(\theta) = \frac{n}{n+1}\left(h(\theta) - \frac{\alpha}{n+1}\right), \quad \theta < \theta < 2\pi, \quad n \geq 1 . \tag{4.1b}$$

Note that $0 = h_o(\theta) < h_1(\theta) < \ldots < h_n(\theta) < h_{n+1}(\theta) < \ldots < h(\theta)$ and $h_n(\theta) \to h(\theta)$ uniformly in $\theta$ as $n \to \infty$. The strict inequalities are essential for our procedure, which is why we assume $\alpha > 0$.

Since we are taking the functions $h_n < h$, we start the iteration with $\bar{u}_o(r)$. By taking $h_n(\theta) = h(\theta) + 1/(n+1)$, $n \geq 0$, and starting the iteration procedure with $\bar{u}_\varepsilon(r)$, we could avoid assuming $\alpha = \min h > 0$ (but still assuming $0 \leq h \leq 1$).

Regarding $\bar{u}_o(r)$ as a function of r and $\theta$, we see that $\bar{u}_o \in C^1(\bar{D}) \cap C^2(D/\Gamma_o)$, where $\Gamma_o = \{(r,\theta): r = r(0)\}$. Let us define the function $U(r,\theta)$ by

$$U(r,\theta) = -\frac{1}{2\pi}\int_D (\ln s) f(\bar{u}_o(t)) + \phi_o(r,\theta) , \tag{4.2}$$

where $r^2 = x^2 + y^2$, $s^2 = (x-\xi)^2 + (y-\eta)^2$, $t = \sqrt{\xi^2+\eta^2}$, and $\phi_o$ is a function harmonic in D chosen such that $U(1,\theta) \equiv 0$. (In integrals over D or some other plane region, the element of area "$d\xi d\eta$" will always be suppressed.)

By Lemma A3 (in the appendix)

$$\Delta U + f(\bar{u}_o) = 0 \qquad \text{in } D/\Gamma_o \tag{4.3a}$$

$$U(1,\theta) \equiv 0, \qquad 0 \leq \theta \leq 2\pi . \tag{4.3b}$$

We already know, however, that $\bar{u}_o$ solves this problem; Theorem 2 in Section 3, on uniqueness, now allows us to conclude that

$$\bar{u}_o(r) = -\frac{1}{2\pi}\int_D (\ln s) f(\bar{u}_o(t)) + \phi_o(r,\theta) . \tag{4.4}$$

The next function in the iteration scheme (1.3) may be represented by

$$u_1(r,\theta) = -\frac{1}{2\pi}\int_D (\ln s) f(\bar{u}_o(t)) + \phi_1(r,\theta) + \varepsilon\psi_1(r,\theta). \tag{4.5}$$

where $\phi_1$ and $\psi_1$ are harmonic in D, $\phi_1$ is such that $-\frac{1}{2\pi}\int_D (\ln s) f(\bar{u}_o(t)) + \phi_1$ vanishes on $r = 1$, and $\psi_1(1,\theta) = h_1(\theta)$, so that $u_1(1,\theta) = \varepsilon h_1(\theta)$. (According to Lemma A3, the function $u_1$ given by (4.5) is a solution of (1.3) with $n = 0$, and it is the only solution, by Theorem 2). Since the integral terms in (4.4) and (4.5) are the same, $\phi_1 \equiv \phi_o$; this result does not hold for further iterates.

We now compare $u_1$ and $u_o$ by using the minimum principle, Theorem 1.

Specifically, since $f(u_1) - f(u_0) \geq 0$,

$$\Delta(u_1 - \bar{u}_0) \leq 0 \qquad \text{in } D/\Gamma_0, \tag{4.6a}$$

$$(u_1 - \bar{u}_0) = \varepsilon h_1 \qquad \text{on } r = 1 . \tag{4.6b}$$

By the definition of the functions $h_n$, $\varepsilon \min h_1 \equiv \mu_1 > 0$, and so the minimum principle implies

$$u_1(r,\theta) - \bar{u}_0(r) \geq \mu_1 > 0 \text{ in } \bar{D}. \tag{4.7}$$

We can also use the minimum principle to conclude that

$$\bar{u}_\varepsilon(r) - u_1(r,\theta) \geq \nu_1 > 0 \text{ in } \bar{D} , \tag{4.8}$$

where $\nu_1 = \varepsilon \min(1-h_1(\theta)) > 0$. We thus have

$$\bar{u}_0(r) + \mu_1 \leq u_1(r,\theta) \leq \bar{u}_\varepsilon(r) - \nu_1, \tag{4.9}$$

$\mu_1 > 0$ and $\nu_1 > 0$.

Consider the annulus

$$D(\varepsilon) = \{(r,\theta)\colon r(0) < r < r(\varepsilon), \qquad 0 \leq \theta \leq 2\pi\}. \tag{4.10}$$

From (4.4) and (4.5)

$$u_1(r,\theta) - \bar{u}_0(r) = \varepsilon\psi_1(r,\theta). \tag{4.11}$$

Since $\psi_1$ is harmonic in D and $0 < \psi_1 \leq 1$ on $\partial D$ (because $0 < h_1 \leq 1$), we see [6, p. 125] that

$$|\partial_x(u_1 - \bar{u}_0)| = \varepsilon|\partial_x\psi_1| \leq \frac{4\varepsilon}{\pi(1-r(\varepsilon))} , \qquad r(0) \leq r \leq r(\varepsilon) .$$

A similar bound holds for $|\partial_y(u_1-\bar{u}_0)|$. Since $\partial_r = \cos\theta\, \partial_x + \sin\theta\, \partial_y$, these bounds give

$$|\partial_r(u_1 - \bar{u}_0)| \leq \frac{8\varepsilon}{\pi(1-r(\varepsilon))} \quad \text{in } D(\varepsilon). \tag{4.12}$$

From Section 2 we know that $|r(\varepsilon) - r(0)| \to 0$ as $\varepsilon \to 0$. If now $\varepsilon$ is chosen small enough that $|r(\varepsilon) - r(0)| < \frac{1}{2}(1-r(0))$, (4.12) yields

$$|\partial_r(u_1 - \bar{u}_0)| \leq \frac{16\varepsilon}{\pi(1-r(0))} \text{ in } D(\varepsilon). \tag{4.12'}$$

Recalling, by (2.3), that

$$\partial_r\bar{u}_0 = -\frac{1}{r}\int_0^r s\, f(\bar{u}_0(s))ds$$

we obtain, for $r(0) \leq r \leq r(\varepsilon)$,

$$\partial_r\bar{u}_0 \leq \frac{-1}{2r(0)}\int_0^{r(0)} s\, ds = -\frac{r(0)}{4} \tag{4.13}$$

Hence, from (4.12') and (4.13)

$$\partial_r u_1 \leq -\frac{r(0)}{4} + \frac{16\varepsilon}{\pi(1-r(0))} < 0 \tag{4.14}$$

for $\varepsilon$ small enough.

For fixed $\theta$, therefore, $u_1$ is strictly decreasing in $r$ for $(r,\theta) \in D(\varepsilon)$. Furthermore, since $\bar{u}_0 = \mu$ for $r = r(0)$ and $\bar{u}_\varepsilon = \mu$ for $r = r(\varepsilon)$, (4.9) implies $u_1(r,\theta) > \mu$ for $r < r(0)$ and $u_1(r,\theta) < \mu$ for $r > r(\varepsilon)$. Hence, there exists a function $\gamma_1(\theta)$ such that $u_1(\gamma_1(\theta),\theta) = \mu$, $0 \leq \theta \leq 2\pi$, with $r(0) < \gamma_1(\theta) < r(\varepsilon)$, and $u_1(r,\theta) \neq \mu$ for $r \neq \gamma_1(\theta)$. Since $\partial_r u_1 < 0$ in $D(\varepsilon)$ and $u_1$ is $C^2$ in $D(\varepsilon)$, the implicit function theorem implies that $\gamma_1(\theta)$ is a $C^2$, $2\pi$-periodic function.

We now estimate $|\gamma_1'(\theta)|$. Since $u_1(\gamma_1(\theta),\theta) = \mu$ ,

$$\frac{d\gamma_1(\theta)}{d\theta} = -\frac{\partial_\theta u_1}{\partial_r u_1} . \tag{4.15}$$

From (4.14), $\partial_r u_1 \leq -\frac{r(0)}{4} + \frac{16\varepsilon}{\pi(1-r(0))} < -\frac{r(0)}{8}$ for $\varepsilon$ small enough.

From (4.5),

$$\partial_\theta u_1 = -\frac{1}{2\pi}\partial_\theta\int_D(\ln s)f(\bar{u}_0(t)) + \partial_\theta\phi_1 + \varepsilon\partial_\theta\psi_1. \tag{4.16}$$

To use $\partial_\theta = -r\sin\theta\,\partial_x + r\cos\theta\,\partial_y$, we estimate x- and y-derivatives for points in $D(\varepsilon)$. Since $0 < \psi_1 \leq 1$ on $\partial D$ and $\psi_1$ is harmonic in D, we can estimate $|\partial_x\psi_1|$ and $|\partial_y\psi_1|$ as before.

If we can bound $\phi_1$ on $\partial D$, we can also bound $|\partial_x\phi_1|$ and $|\partial_y\phi_1|$ in $D(\varepsilon)$. However, we know $\phi_1$ on $\partial D$:

$$\phi_1(1,\theta) = \frac{1}{2\pi}\int_D (\ln s)f(\bar{u}_o(t)), \qquad x^2 + y^2 = 1 . \tag{4.17}$$

For all points in $\bar{D}$ we have, by Lemma A1,

$$\left| -\frac{1}{2\pi}\int_D (\ln s)f(\bar{u}_o(t))\right| < \frac{1}{2\pi}\int_D |\ln s| \leq e^{-1}\delta + |\ln\delta| \tag{4.18}$$

for any $\delta \in (0, 1/2)$. Choosing $\delta = 1/4$, say, gives

$$|\phi_1(1,\theta)| \leq \ln 4 + 1/4e. \tag{4.19}$$

Since $\phi_1$ is harmonic in D, $||\phi_1|| \leq \ln 4 + 1/4e$, where $||\cdot||$ denotes the supremum norm.

Now we proceed as with $\psi_1$ to estimate $|\partial_x\phi_1|$ and $|\partial_y\phi_1|$. The only thing left to estimate is the first term on the right side of (4.16). Using Lemma A1, we find that

$$\left| -\frac{1}{2\pi}\partial_x \int_D (\ln s)f(\bar{u}_o(t))\right| \leq \frac{1}{2\pi}\int_D \frac{1}{s} \leq \delta + \delta^{-1}, \tag{4.20}$$

where $\delta \in (0, 1/2)$. Again choosing $\delta = 1/4$, we have

$$\left| -\frac{1}{2\pi}\partial_x \int_D (\ln s)f(\bar{u}_o(t))\right| \leq 17/4 . \tag{4.21}$$

Combining all these estimates and proceeding as in [6], we obtain

$$|\partial_\theta u_1| \leq 2\left\{\frac{4(17/4)}{\pi(1-r(\varepsilon))}\right\} + 2\left\{\frac{4(\ln 4 + 1/4e)}{\pi(1 - r(\varepsilon))}\right\}$$

$$+ \varepsilon\left\{\frac{4}{\pi(1 - r(\varepsilon))}\right\} < \hat{c}_1 < \infty , \tag{4.22}$$

where $\hat{c}_1$ is a constant independent of $\varepsilon$. Finally, using (4.22) and $\partial_r u_1 \leq -r(0)/8$ in (4.15), we obtain

$$\left|\frac{d\gamma_1(\theta)}{d\theta}\right| \leq \frac{\hat{c}_1}{r(0)/8} \equiv c_1 < \infty. \tag{4.23}$$

We note that the above estimates will hold for all further iterates, since we bounded $f(\bar{u}_0)$ by 1, its maximum value, and an estimate similar to (4.14) can be derived for all other iterates.

We have now completed our study of $u_1(r,\theta)$. The necessary estimates for $u_2$ will be obtained in a way that indicates how all further iterates can be handled. According to Theorem 2 and Lemma A3, the unique solution $u_2$ of problem (1.3), with $n = 1$, may be represented by

$$u_2(r,\theta) = -\frac{1}{2\pi}\int_D (\ln s)\, f(u_1) + \phi_2(r,\theta) + \varepsilon\psi_2(r,\theta), \tag{4.24}$$

where $\phi_2$ and $\psi_2$ are harmonic in D, $-\frac{1}{2\pi}\int_D (\ln s) f(u_1) + \phi_2$ vanishes on $r = 1$, and $\psi_2(1,\theta) = h_2(\theta)$. Also, $u_2 \in C^1(\bar{D}) \cap C^2(D/\Gamma_1)$, $\Gamma_1 = \{(r,\theta):r = \gamma_1(\theta)\}$. Recalling the respective BVP's (1.3) satisfied by $u_1, u_2$ and $\bar{u}_\varepsilon$, we use the minimum principle to obtain

$$u_2(r,\theta) - u_1(r,\theta) \geq \varepsilon \min(h_2 - h_1) = \mu_2 > 0\ , \tag{4.25}$$

$$\bar{u}_\varepsilon(r) - u_2(r,\theta) \geq \varepsilon \min(1 - h_2) = \nu_2 > 0\ . \tag{4.26}$$

Thus, we find

$$u_1(r,\theta) + \mu_2 \leq u_2(r,\theta) \leq \bar{u}_\varepsilon(r) - \nu_2\ . \tag{4.27}$$

We now want to estimate $|\partial_r(u_2 - \bar{u}_0)|$ in $D(\varepsilon)$. As before, $\partial_r = \cos\theta\ \partial_x + \sin\theta\ \partial_y$. We shall find an estimate for

$$\partial_x(u_2 - \bar{u}_0) = -\frac{1}{2\pi}\partial_x \int_D (\ln s)(f(u_1) - f(u_0)) + \partial_x(\phi_2 - \phi_0) + \varepsilon\partial_x\, \psi_2\ , \tag{4.28}$$

that for $\partial_y(u_2 - \bar{u}_0)$ following similarly. First, from Lemma A2,

$$\left| -\frac{1}{2\pi}\partial_x \int_D (\ln s)(f(u_1) - f(\bar{u}_0))\right| \leq \frac{1}{2\pi}\left|\partial_x \int_{D(\varepsilon)} (\ln s)\right| < \delta + \delta^{-1}(r(\varepsilon) - r(0)),$$

for $0 < \delta < 1/2$. We also have

$$|\phi_2 - \phi_o| < \left| -\frac{1}{2\pi}\int_D (\ln s)(f(u_1) - f(\bar{u}_o))\right|$$

$$< \frac{1}{2\pi}\int_{D(\varepsilon)} |\ln s| \leq e^{-1}\delta + |\ln \delta|(r(\varepsilon) - r(0)) \quad ,$$

for $0 < \delta < 1/2$. Because $0 < h_2 \leq 1$, we have $0 < \psi_2 \leq 1$ on $\partial D$. Combining these, we find, for $(r,\theta) \in D(\varepsilon)$,

$$|\partial_r(u_2 - \bar{u}_o)| \leq 2B\{\delta + \delta^{-1}(r(\varepsilon) - r(0))\}$$
$$+ 2B\{e^{-1}\delta + |\ln \delta|(r(\varepsilon) - r(0))\} + 2B\varepsilon, \tag{4.29}$$

where B is defined by

$$B = \frac{4}{\pi(1 - r(\varepsilon))} \; . \tag{4.30}$$

Choosing $\delta$ so that $2B(\delta + e^{-1}\delta) < r(0)/16$, then choosing $\varepsilon$ so that $2B\{(\delta^{-1} + |\ln \delta|)(r(\varepsilon) - r(0)) + \varepsilon\} < r(0)/16$, we get from (4.29)

$$|\partial_r(u_2 - \bar{u}_o)| < r(0)/8, \; (r,\theta) \in D(\varepsilon). \tag{4.31}$$

From (4.13), $\partial_r\bar{u}_o < - r(0)/4$, which means that

$$\partial_r u_2 \leq - r(0)/8 < 0 \quad \text{in } D(\varepsilon). \tag{4.32}$$

We note that the same estimate can be derived for all further iterates, with the same values of $\delta$ and $\varepsilon$.

Since $\partial_r u_2 < 0$ in $D(\varepsilon)$, (4.27) implies the existence of a function $\gamma_2(\theta)$ such that $u_2(\gamma_2(\theta),\theta) = \mu$, $u_2(r,\theta) \neq \mu$ for $(r,\theta) \neq (\gamma_2(\theta),\theta)$, and $\gamma_1(\theta) < \gamma_2(\theta) < r(\varepsilon)$. The implicit function theorem guarantees that $\gamma_2$ is $C^2$, so the minimum principle can be applied at the next stage. We can bound $|d\gamma_2/d\theta|$ by the same constant $c_1$ that we used for $|d\gamma_1/d\theta|$, using the same procedure.

## 5. CONVERGENCE PROOF; EXISTENCE RESULT

The iteration procedure now continues in a similar manner for all further iterates. All necessary estimates are derived in the same way. We thus generate two sequences of functions, $\{u_n(r,\theta)\}$ and $\{\gamma_n(\theta)\}$, such that

$$\bar{u}_0(r) \leq \dots \leq u_n(r,\theta) \leq u_{n+1}(r,\theta) \leq \dots \leq \bar{u}_\varepsilon(r), \tag{5.1}$$

$$r(0) \leq \dots \leq \gamma_n(\theta) \leq \gamma_{n+1}(\theta) \leq \dots \leq r(\varepsilon). \tag{5.2}$$

By virtue of the common estimate $|\gamma_n'(\theta)| \leq c_1 < \infty$ and (5.2), the Ascoli-Arzela theorem implies the existence of a continuous function $\gamma_\infty(\theta)$ such that $\gamma_n \to \gamma_\infty$ uniformly on $[0,2\pi]$. (The entire sequence converges because of the monotonicity.)

Since the sequence of functions $F_n = -\frac{1}{2\pi} \int_D (\ln s)f(u_{n-1})$ is equicontinuous and uniformly bounded (because $\int_D \ln s$ is a continuous function in $\bar{D}$), there is a uniformly convergent subsequence $F_{n_k}$ in $\bar{D}$. The convergence on $\partial D$ implies that the harmonic functions $\phi_{n_k}$ $(= -F_{n_k}$ on $\partial D)$ converge uniformly on $\partial D$. By Harnack's theorem, these functions converge uniformly in $\bar{D}$ to a function $\phi_\infty$, which is harmonic in D. Since $\psi_n(1,\theta) = h_n(\theta) \to h(\theta)$ uniformly for $0 \leq \theta \leq 2\pi$, the harmonic functions $\psi_n$ converge uniformly in $\bar{D}$ to $\psi_\infty$, and $\psi_\infty(1,\theta) = h(\theta)$, $0 \leq \theta \leq 2\pi$. Thus, $\psi_{n_k} \to \psi_\infty$ uniformly in $\bar{D}$. Since $u_{n_k} = F_{n_k} + \phi_{n_k} + \varepsilon\psi_{n_k}$, it follows that $u_{n_k}$ converges uniformly in $\bar{D}$ to a function $u_\infty$. The monotonicity of the sequence $\{u_n\}$ now implies that the entire sequence converges uniformly in $\bar{D}$.

Assuming for the moment that $\int_D (\ln s)f(u_n) \to \int_D (\ln s)f(u_\infty)$ uniformly in $\bar{D}$, we see that the entire sequence $\phi_n$ converges to $\phi_\infty$ on $\partial D$, and so $\phi_n \to \phi_\infty$ uniformly in $\bar{D}$. We now have

$$u_\infty = -\frac{1}{2\pi} \int_D (\ln s)f(u_\infty) + \phi_\infty + \varepsilon\psi_\infty \quad \text{in } \bar{D}\,. \tag{5.3}$$

Since $\phi_\infty$ and $\psi_\infty$ are harmonic and $u_\infty(r,\theta) \neq \mu$ for $(r,\theta) \neq (\gamma_\infty(\theta),\theta)$ (because of the uniform convergence), we have $u_\infty \in C^1(\bar{D}) \cap C^2(D/\Gamma_\infty)$,

$\Gamma_\infty = \{(r,\theta): r = \gamma_\infty(\theta),\ 0 \leq \theta \leq 2\pi\}$. Further, by Lemma A3

$$\Delta u_\infty + f(u_\infty) = 0 \qquad \text{in } D/\Gamma_\infty , \tag{5.4}$$

$$u_\infty(1,\theta) = \varepsilon h(\theta), \qquad 0 \leq \theta \leq 2\pi . \tag{5.5}$$

The $C^1$ character of $\gamma_\infty$ is now established by applying the implicit function theorem to the relation $u_\infty(r,\theta) = \mu$, essentially as done above for $\gamma_i$ $(i = 1, 2, \ldots)$.

The only thing remaining to be proved is the uniform convergence of $\int_D (\ln s) f(u_n)$ to $\int_D (\ln s) f(u_\infty)$. Let $D = D_1 \cup D_N \cup D_2$, where

$$D_1 = \{(r,\theta): 0 \leq r \leq \gamma_N(\theta),\ 0 \leq \theta \leq 2\pi\} , \tag{5.6a}$$

$$D_N = \{(r,\theta): \gamma_N(\theta) \leq r \leq \gamma_\infty(\theta), 0 \leq \theta \leq 2\pi\}, \tag{5.6b}$$

$$D_2 = \{(r,\theta): \gamma_\infty(\theta) \leq r \leq 1,\ 0 \leq \theta \leq 2\pi\} . \tag{5.6c}$$

On $D_1 \cup D_2$ we have $f(u_\infty) - f(u_n) = 0$ for $n \geq N$. Thus,

$$\left| \int_D (\ln s)(f(u_\infty) - f(u_n)) \right| \leq \int_{D_N} |\ln s|$$

$$\leq 2\pi \{e^{-1}\delta + |\ln \delta|\ \|\gamma_\infty - \gamma_N\|\},\quad 0 < \delta < 1/2,\ n \geq N.$$

Given $\nu > 0$, choose $\delta < \nu/(8\pi e^{-1})$ and then choose N so that $\|\gamma_\infty - \gamma_N\| < \nu/(8\pi|\ln \delta|)$. Now, $n \geq N$ implies $|\int_D (\ln s)(f(u_\infty) - f(u_n)| < \nu$ for $n \geq N$ for all $(x,y) \in \bar{D}$. Thus, we have the desired uniform convergence.

We have proved the following existence result.

<u>Theorem 3</u>. Suppose $\mu \in (0, 1/4e)$. For $\varepsilon$ small enough, the sequence $\{u_n(r,\theta)\}$ converges monotonically and uniformly to a limit function $u_\infty(r,\theta)$, where $\bar{u}_0(r) \leq u_\infty(r,\theta) \leq \bar{u}_\varepsilon(r)$, and the sequence $\{\gamma_n(\theta)\}$ to $\gamma_\infty(\theta)$, which is $C^1$ and $2\pi$-periodic and satisfies $r(0) \leq \gamma_\infty(\theta) \leq r(\varepsilon)$ for $0 \leq \theta \leq 2\pi$. Let $\Gamma_\infty = \{(r,\theta): r = \gamma_\infty(\theta)\}$. Then $u_\infty(r,\theta)$ is a solution of BVP (1.1), with free boundary $\Gamma_\infty$.

## 6. REMARKS

We have established the existence of a classical solution to a nonlinear Poisson interface problem within the conical segment $<\bar{u}_o(r), \bar{u}_\varepsilon(r)>$. Solving Laplace's equation in D with $u = \varepsilon h(\theta)$ on $\partial D$ provides another solution, $\underline{u}$, contained in $<\underline{u}_o(r), \underline{u}_\varepsilon(r)> = <0,\varepsilon>$, provided $\varepsilon < \mu$. This solution is less interesting since it avoids the nonlinearity due to the function f. The function $f(u_\infty)$ takes on both values, zero and one, whereas $f(\underline{u}) \equiv 0$ in $\bar{D}$.

Similar techniques can be applied to more complicated problems. In particular, if we take the nonlinear term to have the form

$$f = f(r,u) = \begin{cases} f_1(r,u), & u < \mu \\ f_2(r,u), & \mu < u \ , \end{cases}$$

a completely analogous procedure provides the same sort of existence theorem.

The existence result also provides some justification for a perturbation (or linearization) procedure developed by the authors [5]. In Section 2 we saw that $r(\varepsilon) - r(0) = O(\varepsilon)$ and $\bar{u}_\varepsilon(r) - \bar{u}_o(r) = O(\varepsilon)$, while the existence theorem of Section 5 showed that $r(0) \le \gamma_\infty(\theta) \le r(\varepsilon)$ and $\bar{u}_o(r) \le u_\infty(r,\theta) \le \bar{u}_\varepsilon(r)$. We thus have the result that $\gamma_\infty(\theta) = r(0) + O(\varepsilon)$ and $u_\infty(r,\theta) = \bar{u}_o(r) + O(\varepsilon)$, as assumed in the perturbation procedure.

## APPENDIX

We prove three results used previously.

Lemma A1. If A is an open subset of D,

$$\int_A |\ln s| \le 2\pi e^{-1}\delta + |\ln \delta|(\text{area of A}) \quad , \ 0 < \delta < 1/2.$$

Proof: For $(x,y) \in \bar{D}$ and $\delta \in (0, 1/2)$, let N be the disc of radius $\delta$ centered on $(x,y)$. Since $A = A \cap N \cup A/N$,

$$\begin{aligned} \int_A |\ln s| &= \int_{A\cap N} |\ln s| + \int_{A/N} |\ln s| \\ &\le \int_N |\ln s| + |\ln \delta| \int_A 1 \\ &\le 2\pi\delta e^{-1} + |\ln \delta|(\text{area of A}) \ , \end{aligned}$$

where we have used the fact that $0 \le s \le 2$ in $\bar{D}$, and transformed to polar coordinates to estimate the integral over N.

Lemma A2. Let A be as above; then similar procedures yield

$$\int_A \frac{1}{s} \le 2\pi\delta + \delta^{-1}(\text{area of } A), \qquad 0 < \delta < 1/2 .$$

Lemma A3. Let $\Gamma = \{(r,\theta): r = \gamma(\theta),\ 0 \le \theta \le 2\pi\}$, where $\gamma$ is continuous, $2\pi$-periodic and satisfies $0 < \gamma(\theta) < 1$. If $g(x,y) \in C^1(\bar{D}/\Gamma)$, then the function v defined by

$$v(x,y) = \int_D (\ln s) g(\xi,\eta)$$

belongs to $C^1(\bar{D}) \cap C^2(D/\Gamma)$ and satisfies

$$\Delta v = 2\pi g(x,y) \quad \text{in } D/\Gamma.$$

Proof: That $v \in C^1(\bar{D})$ follows from the argument used in [6, pp. 151-153], which is still applicable under the weaker assumption on g made here. Given $(x_o, y_o)$ in $D/\Gamma$, let N be a disc of radius $\delta$ centered on $(x_o, y_o)$ and with $\bar{N} \subset D/\Gamma$. Then

$$v(x,y) = \int_N (\ln s)g + \int_{D/N} (\ln s)g \equiv I_1(x,y) + I_2(x,y).$$

For $(x,y)$ within $\delta/2$ of $(x_o, y_o)$, $I_2(x,y) \in C^\infty$ and is harmonic. As in [6], $I_1 \in C^2(N)$ and satisfies $\Delta I_1 = 2\pi g$ since $g \in C^1(\bar{N})$. Thus, $v \in C^2$ for $(x,y)$ in a neighborhood of radius $\delta/2$ of $(x_o, y_o)$, and satisfies $\Delta v = 2\pi g$ there. Since $(x_o, y_o) \in D/\Gamma$ was arbitrary, the result follows.

REFERENCES

1. Amann, H., "Order structures and fixed points," lecture notes, Ruhr-Universität, Bochum 1977.

2. Chandra, J., and Fleishman, B.A., "Existence and comparison results for a class of nonlinear boundary value problems," Annali di Matematica Pura ed Applicata, Serie IV-Tomo CI(1974), 247-261.

3. Fleishman, B.A., and Mahar, T.J., "Boundary value problems for a nonlinear differential equation with discontinuous nonlinearities," Math. Balk. 3 (1973), 98-108.

4. Fleishman, B.A., and Mahar, T.J., "Boundary value problems with discontinuous nonlinearities: comparison of solutions, approximation, and continuous dependence on parameters," J. Diff. Eq. 26 (1977), 262-277.

5. Fleishman, B.A., and Mahar, T.J., "Analytic methods for approximate solution of elliptic free boundary problems," Nonlinear Analysis: Theory, Methods, and Applications 1 (1977), 561-569.

6. John, F., "Partial Differential Equations," Appl. Math. Sciences 1, Springer-Verlag, New York, 1971 .

7. Protter, M.H., and Weinberger, H.F., "Maximum principles in differential equations," Prentice-Hall, Inc., Englewood Cliffs, N.J., 1967 .

*Differential Equations and Applications*
*W. Eckhaus and E.M. de Jager (eds.)*
*©North-Holland Publishing Company (1978)*

# ASYMPTOTIC SOLUTIONS IN FREE BOUNDARY PROBLEMS OF SINGULARLY PERTURBED ELLIPTIC VARIATIONAL INEQUALITIES

W. Eckhaus and H.J.K. Moet
Mathematisch Instituut
der Rijksuniversiteit Utrecht

The theory of elliptic variational inequalities, which is due to Stampacchia [12] and Lions and Stampacchia [10], is modelled after the variational theory of elliptic boundary value problems, however, its scope is much wider. See, for instance, Duvaut and Lions [2] and Lions [7] and the references therein.

As is well known [9] the variational theory of elliptic boundary value problems leads in a natural way to an elliptic variational problem, solutions of which are called weak solutions of the corresponding boundary value problem. It is interesting to note that an elliptic variational inequality of unilateral type (see example 1 of [10]) can sometimes be translated into a "differential inequality", such that the (unique) solution of the variational inequality is a solution, in a distributional sense, of this "differential inequality"; cf. [10] for details.

It is tempting to attack these differential inequalities directly and it is even more tempting to attack the singularly perturbed differential inequalities as formulated by Lions in his Scheveningen lectures [8]. In the present paper we develop a method of analysis for a certain class of singularly perturbed elliptic variational inequalities via associated differential inequalities. We shall show that, by using the knowledge of the behavior of solutions of singularly perturbed elliptic boundary value problems, it is possible to "determine" a priori the free boundary (to be defined shortly) and then approximate the solution by standard techniques of asymptotic analysis of singular perturbations, following Eckhaus and de Jager [6]. To define the ideas we recall the formulation of the problem and some basic results.

Let $\Omega$ be a bounded open set in $\mathbb{R}^n$ with a smooth boundary $\partial\Omega$. We consider the elliptic operator $L_\varepsilon$, given by

$$(1.1)\qquad L_\varepsilon = -\varepsilon \sum_{i,j=1}^{n} \frac{\partial}{\partial x_i}\left(a_{ij}(x)\frac{\partial}{\partial x_j}\right) + \sum_{i=1}^{n} a_i(x)\frac{\partial}{\partial x_i} + a_0(x),$$

where $\varepsilon$ is a small positive parameter and $a_{ij}, a_i, a_0$ are real valued functions belonging to $L^\infty(\Omega)$.

Let $H^1(\Omega)$ denote as usual the Sobolev space of functions which, together with their first derivatives belong to $L^2(\Omega)$, and let $H_0^1(\Omega)$ denote the closure of $C_0^\infty(\Omega)$ in $H_1(\Omega)$.

We associate to the differential operator $L_\varepsilon$ the bilinear form $a(u,v), u,v \in H^1(\Omega)$, defined by

$$(1.2)\quad a(u,v) = \varepsilon \sum_{i,j=1}^{n} \int_\Omega a_{ij}(x)\frac{\partial u}{\partial x_i}\frac{\partial v}{\partial x_j}dx + \sum_{i=1}^{n}\int_\Omega a_i(x)\frac{\partial u}{\partial x_i}v dx + \int_\Omega a_0(x)uv dx.$$

We now consider the following problem (a variational inequality):

<u>Determine an element</u> $u \in H_0^1(\Omega)$, $u \leqslant 0$ <u>in</u> $\Omega$, <u>such that for each</u> $v \in H_0^1(\Omega)$, $v \leqslant 0$ in $\Omega$, <u>one has</u>

$$(1.3)\quad a(u,v-u) \geqslant (f,v-u),$$

<u>where</u> f <u>is a given element of</u> $L^2(\Omega)$ <u>and</u> $(.,.)$ <u>indicates the usual inner product in</u> $L^2(\Omega)$.

Suppose that the biliner form $a(.,.)$ is such that, for any $u,v \in H_0^1(\Omega)$,

$$(1.4)\quad |a(u,v)| \leqslant C\|u\|_{H^1(\Omega)}\|v\|_{H^1(\Omega)} \text{ for some constant } C \text{ (continuity)}$$

$$(1.5)\quad a(v,v) \geqslant \alpha\|v\|^2_{H^1(\Omega)} \quad \text{for some } \alpha > 0 \quad \text{(coerciveness)}$$

then, by a fundamental result of Stampacchia ([10], [12]), the problem formulated above possesses a unique solution. Furthermore, it can be shown that, if the solution is sufficiently smooth, i.e., $u \in H^2(\Omega)$, it satisfies the following set of relations:

$$(1.6)\quad L_\varepsilon u - f \leqslant 0 \quad \text{in } \Omega$$

$$(1.7)\quad u/\partial\Omega = 0 \text{ and } u \leqslant 0 \quad \text{in } \Omega$$

$$(1.8)\quad (L_\varepsilon u - f)u = 0 \quad \text{in } \Omega$$

These relations suggest that the solution u of the variational inequality is a function such that in some set $\Omega_0 \subset \Omega$ one has $L_\varepsilon u - f = 0$, while in some other set $\Omega_1 \subset \Omega$ the function u is identically zero. We shall call the boundary $\partial\Omega_0$ of $\Omega_0$ the free boundary. The above set of relations will be referred to as a differential inequality.

It is easily seen that the differential inequality is not necessarily uniquely solvable, in $H_0^1(\Omega)$, however, if we add to the above set of relations (1.6), (1.7), (1.8), one further requirement (which will be specified in due course) the problem has a unique solution and for a certain class of differential operators $L_\varepsilon$ this solution satisfies the corresponding variational inequality.

In order to bring out clearly the ideas and the method of analysis, we study in this paper some simple but representative problems in $\mathbb{R}^1$. More detailed results for problems of greater complexity will be given in a subsequent publication. We commence by deriving some elementary results on the behavior of solutions.

## 2. ELEMENTARY PROPERTIES OF SOLUTIONS

In all that follows $\Omega$ will stand for an open, bounded subset of $\mathbb{R}^n$ $(n \geq 1)$ and we assume that the following conditions are satisfied:

(2.1) the coefficients $a_{ij}, a_i, a_0$ of the differential operator $L_\varepsilon$, given by

$$L_\varepsilon = -\varepsilon \sum_{i,j=1}^{n} \frac{\partial}{\partial x_i}\left(a_{ij}(x)\frac{\partial}{\partial x_j}\right) + \sum_{i=1}^{n} a_i(x)\frac{\partial}{\partial x_i} + a_0(x),$$

are real valued and all belong to $C(\overline{\Omega}) \cap C^\infty(\Omega)$;

(2.2) $L_\varepsilon$ is uniformly elliptic in $\Omega$, i.e., there exists a constant $A > 0$ such that, for all $x \in \Omega$,

$$\sum_{i,j=1}^{n} a_{ij}(x)\,\xi_i\xi_j \geq A \sum_{i=1}^{n} \xi_i^2 \quad \text{for all } \xi \in \mathbb{R}^n;$$

(2.3) the bilinear form induced by $L_\varepsilon$

$$a(u,v) = \varepsilon \sum_{i,j=1}^{n} \int_\Omega a_{ij}(x)\frac{\partial u}{\partial x_i}\frac{\partial v}{\partial x_j}dx + \sum_{i=1}^{n}\int_\Omega a_i(x)\frac{\partial u}{\partial x_i}v\,dx + \int_\Omega a_0(x)uv\,dx$$

is such that, for any $u,v \in H_0^1(\Omega)$,

$|a(u,v)| \leq C\|u\|_{H^1(\Omega)}\|v\|_{H^1(\Omega)}$ for some constant $C > 0$ independent of $\varepsilon$

$a(v,v) \geq \alpha\|v\|^2_{H^1(\Omega)}$ for some constant $\alpha > 0$ independent of $\varepsilon$;

(2.4) $a_0(x) \geq 0$ for every $x \in \Omega$;

(2.5) the function $f$ belongs to $C(\overline{\Omega}) \cap C^\infty(\Omega)$.

We shall now derive a useful property of the solution $u \in H_0^1(\Omega)$, $u \leq 0$ in $\Omega$, of the variational inequality

(2.6) $a(u,v-u) \geq (f,v-u)$ for each $v \in H_0^1(\Omega)$, $v \leq 0$ in $\Omega$.

Lemma 1. Let u be the unique solution of the variational inequality (2.6). Then u cannot be identically zero in any open subset $\Omega_-^*$ of $\Omega = \{x \in \Omega/f(x) < 0\}$. If for some open subset $\Omega_+^*$ of $\Omega_+ = \{x \in \Omega/f(x) > 0\}$ u belongs to $C^2(\Omega_+^*)$ and $u/\partial\Omega_+^* = 0$, then u must be zero everywhere in $\Omega_+^*$.

Proof. It is a simple matter to verify that (2.6) is equivalent to

(2.7) $a(u,v) \geq (f,v)$ for each $v \in H_0^1(\Omega)$, $v \leq 0$ in $\Omega$

(2.8) $a(u,u) = (f,u)$ $u \in H_0^1(\Omega)$ and $u \leq 0$ in $\Omega$.

Suppose that $u \in H_0^1(\Omega)$ and $u = 0$ in $\Omega_-^*$, then, because $u \in H^1(\Omega_-^*)$, we have $a(u,v) = 0$ for every $v \in C_0^\infty(\Omega_-^*)$, $v \leq 0$ in $\Omega_-^*$, while $(f,v) = \int_\Omega fv dx > 0$ for $v \in C_0^\infty(\Omega_-^*)$, $v \leq 0$ and $v \not\equiv 0$ in $\Omega_-^*$. This contradicts (2.7), hence u is not identically zero in $\Omega_-^*$.

In order to prove the second part of this lemma we proceed as follows. Let $u \in C^2(\Omega_+^*)$ and $u/\partial\Omega_+^* = 0$ for some open subset $\Omega_+^*$ of $\Omega_+$, then u satisfies

$$\begin{aligned} &L_\varepsilon u - f \leq 0 && \text{in } \Omega_+^* \\ &u/\partial\Omega_+ = 0 \text{ and } u \leq 0 && \text{in } \Omega_+^* \\ &(L_\varepsilon u - f)u = 0 && \text{in } \Omega_+^* \end{aligned}$$

Suppose u is not identically zero in $\Omega_+$, then $U = \{x \in \Omega/u(x) < 0\} \cap \Omega_+^*$ is a nonempty open set. Thus we have the following problem for the restriction of u to U (which we also call u):

$$\begin{aligned} L_\varepsilon u - f &= 0 \text{ in } U \\ u &= 0 \text{ in } \partial U. \end{aligned}$$

As a consequence of the maximum principle (see Protter and Weinberger [11]) $u \geq 0$ everywhere in U. However, since $u \leq 0$ we conclude $u \equiv 0$ in $\Omega_+^*$, which contradicts our previous assumption.

The following lemma is an easy consequence of the maximum principle and a result of Brézis and Stampacchia [1].

Lemma 2. Let u be the solution of the variational inequality (2.6) corresponding to the operator

$$L_\varepsilon = -\varepsilon \sum_{i,j=1}^{n} \frac{\partial}{\partial x_i}\left(a_{ij}(x)\frac{\partial}{\partial x_j}\right) + \sum_{i=1}^{n} a_i(x)\frac{\partial}{\partial x_i} + a_0(x),$$

and let f be such that $\Omega_- = \{x \in \Omega / f(x) < 0\}$ is nonempty. If, in addition to the previous assumptions, the coefficients $a_{ij}$ also belong to $C^1(\bar{\Omega})$, then $u < 0$ throughout $\Omega_-$. Moreover, u is uniformly bounded in $\Omega$ with respect to sufficiently small values of $\varepsilon$.
The solution u attains its minimum at a point $x^0 \in \Omega_- \cup \partial\Omega_-$ and

$$a_0(x^0)u(x^0) \geqslant f(x^0),$$

provided $\Omega = \Omega_- \cup \partial\Omega_- \cup \Omega_+$. No nonzero local minima exist outside $\Omega_- \cup \partial\Omega_-$. Furthermore, if $a_0$ is positive on $\partial\Omega_-$, then $x^0 \in \Omega_-$ and no nonzero local minima exist outside $\Omega_-$.

Remark. We define $\Omega_0 = \{x \in \Omega / f(x) = 0\}$. It is easily seen that u attains its minimum at $x^0 \in \Omega_- \cup \Omega_0$ and no nonzero local minima exist outside $\Omega_- \in \Omega_0$. If $a_0$ is positive on $\Omega_0$, then $x^0 \in \Omega_-$ and no nonzero local minima exist outside $\Omega_-$.

Lemma 2*. Let u be the solution of the variational inequality (2.6) corresponding to the operator

$$L_\varepsilon = -\varepsilon \sum_{i,j=1}^{n} \frac{\partial}{\partial x_i}\left(a_{ij}(x)\frac{\partial}{\partial x_j}\right) + \sum_{i=1}^{n} a_i(x)\frac{\partial}{\partial x_i} + a_0(x),$$

and let f be such that $\Omega_- = \{x \in \Omega / f(x) < 0\}$ is nonempty.

If $\Omega_-$ has the ball property *) and if, in addition to the previous assumptions, the coefficients $a_{ij}$ also belong to $C^1(\bar{\Omega})$, then $u < 0$ throughout $\Omega_- \cup \partial\Omega_-$.

In the remaining part of this section we confine ourselves to elliptic differential operators $L_\varepsilon$ with constant coefficients. There is, of course, no loss of generality in supposing $L_\varepsilon$ has the following form

$$(2.9)\quad L_\varepsilon = -\varepsilon\Delta + \sum_{i=1}^{n} a_i \frac{\partial}{\partial x_i} + a_0,$$

where $\Delta = \sum_{i=1}^{n} \frac{\partial^2}{\partial x_i^2}$ and $a_0 \geqslant 0$. In this case, indeed, the conditions (2.1), (2.2), (2.3), (2.4) are satisfied for $\varepsilon > 0$ small enough.

It has been shown in [1] that the solution u to (2.6) belongs to $H^2(\Omega)$. In particular, if $f \in L^p(\Omega)$ $(1 < p < \infty)$ and $p > n$, then u is an element of $C^{1,\alpha}(\bar{\Omega})$ with $\alpha = 1 - \frac{n}{p}$.

If some $a_i$ is nonzero, then the bilinear form associated to $L_\varepsilon$ is nonsymmetric and it is well known (Lions [7]) that in that case the problem of a variational inequality is not a problem of the Calculus of Variations. We shall show (lemma 3 below)

---

*) An open set $\Omega_-$ has the ball property if and only if for each $x \in \partial\Omega_-$ there exists an open ball $B \subset \Omega_-$ such that $\bar{B} \cap \bar{\Omega}_- \ni x$.

that in the case of constant coefficients, we can nevertheless associate to the variational inequality a problem of the Calculus of Variations. This is essentially accomplished by the Liouville transformation.

It is a matter of straightforward computation to show that, if we put $w(x) = u(x)\exp\{-\frac{1}{2\varepsilon}\sum_{j=1}^{n} a_j x_j\}$, problem (2.6) is equivalent to finding a function $w \in H_0^1(\Omega) \cap H^2(\Omega)$, $w \leq 0$ in $\Omega$ such that

(2.10) $\tilde{a}(w,v) \geq (f^*,v)$ for each $v \in H_0^1(\Omega)$, $v \leq 0$ in $\Omega$

(2.11) $(\tilde{L}_\varepsilon w - f^*)\, w = 0$ in $\Omega$,

where $\tilde{L}_\varepsilon = -\varepsilon\Delta + (a_0 + \frac{1}{4\varepsilon}\sum_{j=1}^{n} a_j^2)$, $\tilde{a}$ is the bilinear form induced by $\tilde{L}_\varepsilon$ and $f^*(x) = f(x)\exp\{-\frac{1}{2\varepsilon}\sum_{j=1}^{n} a_j x_j\}$.

Since the bilinear form $\tilde{a}$ is symmetric, the problem of finding w is a problem in the Calculus of Variations (see [7]) :

(2.12) $$\begin{cases} \text{Find } w \in H_0^1(\Omega),\ w \leq 0 \text{ in } \Omega \text{ such that } w \text{ minimizes} \\ \tilde{J}(v) = \tfrac{1}{2}\tilde{a}(v,v) - (f^*,v) \quad \text{for all } v \in H_0^1(\Omega), v \leq 0 \text{ in } \Omega. \end{cases}$$

Transforming $w(x) = u(x)\exp\{-\frac{1}{2\varepsilon}\sum_{j=1}^{n} a_j x_j\}$ yields the following problem for u:

(2.13) $$\begin{cases} \text{Find } u \in H_0^1(\Omega),\ u \leq 0 \text{ in } \Omega \text{ such that } u \text{ minimizes} \\ J(v) = \tfrac{1}{2}\tilde{a}(\theta^{-1}v, \theta^{-1}v) - (\theta^{-1}f, \theta^{-1}v) \text{ for all} \\ v \leq H_0^1(\Omega),\ v \leq 0 \text{ in } \Omega, \text{ where } \theta(x) = \exp\{\frac{1}{2\varepsilon}\sum_{j=1}^{n} a_j x_j\}. \end{cases}$$

Of course, these problems are equivalent, modulo the transform $u = \theta w$, and u solves problem (2.6). Since u belongs to $H^2(\Omega) \cap C^1(\bar{\Omega})$, it is possible to transform the last formulation of the problem for u into an equivalent formulation which is better suited for our purpose.

It is easily verified that the solution u of (2.13) satisfies,

$$L_\varepsilon u - f \leq 0 \quad \text{in } \Omega$$
$$(L_\varepsilon u - f)\, u = 0 \quad \text{in } \Omega,$$

hence, there is no loss of generality in minimizing J in the set $\mathcal{U}$, which is specified as follows:

<u>Definition</u>. v is an element of $\mathcal{U}$ if and only if $v \in H_0^1(\Omega)$, v is continuous and piecewise continuously differentiable, $v \leq 0$ in $\Omega$ and furthermore

$$L_\varepsilon v - f \leqslant 0 \quad \text{in } \Omega$$
$$(L_\varepsilon v - f)v = 0 \quad \text{in } \Omega.$$

One easily verifies that for each $v \in \mathcal{U}$ one has $J(v) = -\frac{1}{2} \int_\Omega vf\,\theta^{-2}\,dx$. Thus we arrive at

Lemma 3. Let u be the unique solution of the variational inequality (2.6) corresponding to the operator $L_\varepsilon$ of (2.9). Then $u \in \mathcal{U}$ and

(2.14) $J(u) = \inf\{-\frac{1}{2}\int_\Omega vf\theta^{-2}dx \,|\, v \in \mathcal{U}\}$.

Remark. In the case of ordinary differential operators there is no need to restrict oneself to the case of constant coefficients, because the Liouville transformation then always exists. Thus, in the one-dimensional case, the equivalent of lemma 3 (with a slightly modified function $\theta$) always holds. This problem will be studied seperately in section 4.

## 3. Problems in $\mathbb{R}^1$. The case of symmetric bilinearforms.

Consider the simplest possible problem: the one-dimensional situation $\Omega = (0,1)$, the differential operator $L_\varepsilon = -\varepsilon \frac{d^2}{dx^2} + a_0(x)$, $a_0 \in C^2(\bar{\Omega})$, $a_0(x) \geqslant d > 0$ in $\Omega$ and a smooth function f, i.e. $f \in C^2(\bar{\Omega})$, as sketched in figure 1.

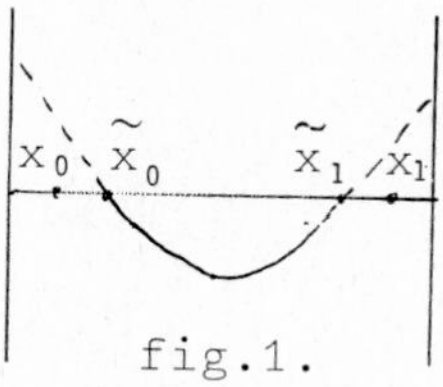

fig.1.

Using the results of section 2 it is a simple matter to verify that the solution u of the variational inequality (2.6) satisfies

(3.1) $L_\varepsilon u - f = 0$ on some interval $(x_0, x_1) \supset (\tilde{x}_0, \tilde{x}_1)$

(3.2) $u = 0$ elsewhere in $\Omega$.

Therefore we may take as competing functions the set $\mathcal{F}$ of all continuous, piecewise twice continuously differentiable functions v satisfying the following conditions:

(3.4) $v \leqslant 0$ in $\Omega$,

(3.5) $L_\varepsilon v - f = 0$ in $(x_0, x_1)$, where $0 \leqslant x < \tilde{x}_0$ and $\tilde{x}_1 < x_1 \leqslant 1$,

(3.6) $v = 0$ in $[0,x_0) \cup (x_1,1]$.

By standard methods of asymptotic analysis of singularly perturbed boundary value problems (see, for instance, Eckhaus[4]) formal approximations for a function v satisfying (3.4), (3.5), (3.6) are easiliy constructed:

(3.7) $$v_{as}(x) = \frac{f(x)}{a_0(x)} - \frac{f(x_0)}{a_0(x_0)} \exp\left(-\sqrt{a_0(x_0)}\,\frac{x-x_0}{\sqrt{\varepsilon}}\right) + -\frac{f(x_1)}{a_0(x_1)} \exp\left(-\sqrt{a_0(x_1)}\,\frac{x_1-x}{\sqrt{\varepsilon}}\right) + R(x_0,x_1,\varepsilon),$$

if $x \in [x_0,x_1]$, and

(3.8) $v_{as}(x) = 0$ if $x \in [0,x_0) \cup (x_1,1]$,

where

(3.9) $$R(x_0,x_1,\varepsilon) = \frac{x-x_0}{x_1-x_0}\,\frac{f(x_0)}{a_0(x_0)} \exp\left(-\sqrt{a_0(x_0)}\,\frac{x_1-x_0}{\sqrt{\varepsilon}}\right) + +\frac{x_1-x}{x_1-x_0}\,\frac{f(x_1)}{a_0(x_1)} \exp\left(-\sqrt{a_0(x_1)}\,\frac{x_1-x_0}{\sqrt{\varepsilon}}\right).$$

In (3.7) we recognize a formal asymptotic approximation of the solution of the following Dirichlet problem:

$$L_\varepsilon \bar{v} - f = 0 \quad \text{in } (x_0,x_1),$$
$$\bar{v}(x_0) = \bar{v}(x_1) = 0,$$

with, however, boundary points $x_0, x_1$ possibly depending on $\varepsilon$.

Fortunately, a proof that the formal approximation given by (3.7), (3.8) is, in fact, an approximation of v in the supremum norm may easily be obtained by a slight modification of standard techniques (Eckhaus[3], Eckhaus and de Jager[6]). As in the case of an $\varepsilon$-independent boundary one finds

(3.10) $v = v_{as} + O(\sqrt{\varepsilon})$,

where the order of magnitude is in the supremum norm. Higher order approximations may also be constructed.

We observe that the condition $v \leqslant 0$ immediately imposes conditions on the position of the endpoints $x_0(\varepsilon)$ and $x_1(\varepsilon)$. Since $f(x) > 0$ for $x < x_0(\varepsilon)$ and $x > x_1(\varepsilon)$, the function v as given by (3.10) will take positive values unless

(3.11) $x_0(\varepsilon) - \tilde{x}_0 = O(\sqrt{\varepsilon})$

(3.12) $x_1(\varepsilon) - \tilde{x}_1 = O(\sqrt{\varepsilon})$.

For assume there is a real $d > 0$ such that for any $\varepsilon_0 > 0$ we can find a positive $\varepsilon < \varepsilon_0$, such that $\tilde{x}_0 - x_0(\varepsilon) \geqslant d$, then there exists a sequence $(\varepsilon_n)$ converging to zero for which $\tilde{x}_0 - x_0(\varepsilon_n) \geq d$. Consider now

$$v(\tilde{x}_0-\tfrac{1}{2}d) = \frac{f(\tilde{x}_0-\frac{1}{2}d)}{a_0(\tilde{x}_0-\frac{1}{2}d)} +$$

$$- \frac{f(x_0)}{a_0(x_0)} \exp(-\sqrt{a_0(x_0)}\,\frac{\tilde{x}_0-\frac{1}{2}d-x_0}{\sqrt{\varepsilon_n}}) + O(\sqrt{\varepsilon_n}).$$

Letting $n \to \infty$, we find $v(\tilde{x}_0-\frac{1}{2}d)$ tending to $\frac{f(\tilde{x}_0-\frac{1}{2}d)}{a_0(\tilde{x}_0-\frac{1}{2}d)}$. This contradicts $v \leq 0$, hence $(x_0(\varepsilon)-\tilde{x}_0) \to 0$ as $\varepsilon \to 0$. Evidently, $(x_1(\varepsilon)-\tilde{x}_1) \to 0$ as $\varepsilon \to 0$ is proved in an analogous way.

To complete the proof of (3.11), (3.12) we investigate the function v in the local variable $\xi = (x-x_0)/\sqrt{\varepsilon}$ (see Eckhaus [4] for terminology and notational conventions):

$$(3.13)\quad a_0(x_0+\xi\sqrt{\varepsilon})v^*(\xi,\varepsilon) = f(x_0+\xi\sqrt{\varepsilon}) +$$

$$- f(x_0)\,\frac{a_0(x_0+\xi\sqrt{\varepsilon})}{a_0(x_0)} \exp(-\sqrt{a_0(x_0)}\xi) + O(\sqrt{\varepsilon}).$$

We already know that $\tilde{x}_0 - x_0(\varepsilon) = \delta(\varepsilon)$, $\delta(\varepsilon) = o(1)$ as $\varepsilon \to 0$, hence by Taylor's theorem

$$f(x_0) = f(\tilde{x}_0+x_0-\tilde{x}_0) = f'(\tilde{x}_0)(x_0-\tilde{x}_0) + O(\delta^2)$$

$$f(x_0+\xi\sqrt{\varepsilon}) = f(\tilde{x}_0+x_0-\tilde{x}_0+\xi\sqrt{\varepsilon}) = f'(\tilde{x}_0)(x_0-\tilde{x}_0+\xi\sqrt{\varepsilon})+O((\delta+\sqrt{\varepsilon})^2).$$

Substitution in (3.13) gives

$$a_0(x_0+\xi\sqrt{\varepsilon})v^*(\xi,\varepsilon) =$$

$$= f'(\tilde{x}_0)\{(x_0-\tilde{x}_0)(1-\frac{a_0(x_0+\xi\sqrt{\varepsilon})}{a_0(x_0)} \exp(-\sqrt{a_0(x_0)\xi}) + \xi\sqrt{\varepsilon}\} +$$

$$+ O(\sqrt{\varepsilon}) + O(\delta^2) + O((\delta+\sqrt{\varepsilon})^2).$$

Suppose now that $\tilde{x}_0 - x_0(\varepsilon) = \delta(\varepsilon)$, such that $\sqrt{\varepsilon} = o(\delta(\varepsilon))$ (note that $\delta > 0$ by lemma 2 of section 2, and $f'(\tilde{x}_0) < 0$). Then for $\varepsilon$ positive and small enough $v^*(\xi,\varepsilon)$ becomes positive, which contradicts $v \leq 0$. Hence, we find $\delta(\varepsilon) = O(\sqrt{\varepsilon})$ and so, since also $u \in \mathfrak{T}$, in the first approximation the free boundary coincides with the zero's of f.

We shall not pursue here the analysis of higher approximations.

Let us finally define the function $u_0$ as follows

$$u_0(x) = \begin{cases} \dfrac{f(x)}{a_0(x)} & \text{if } x \in [\tilde{x}_0,\tilde{x}_1] \\ 0 & \text{if } x \in [0,\tilde{x}_0) \cup (\tilde{x}_1,1]. \end{cases}$$

Given the preceding results, it is a matter of straightforward computation to show that, for any $v \in \mathcal{G}$

$$(v-u_0) \to 0 \text{ in } L^2(\Omega) \text{ as } \varepsilon \to 0.$$

Since the solution u of the variational inequality (2.6) corresponding to the differential operator $L_\varepsilon = -\varepsilon \frac{d^2}{dx^2} + a_0(x)$, is an element of $\mathcal{G}$, we find that

$$(u - u_0) \to 0 \text{ in } L^2(\Omega) \text{ as } \varepsilon \to 0.$$

It should be clear that the results of this section can immediately be extended to more general functions f having any finite number of zero's in the interval $[0,1]$.

We now collect the results of this section in the following

Theorem 1. Let u be the solution of the variational inequality (2.6) induced by the differential operator $L_\varepsilon$ in $\Omega = (0,1)$, defined by

$$L_\varepsilon = -\varepsilon \frac{d^2}{dx^2} + a_0(x),$$

where $a_0 \in C^2(\bar{\Omega})$, $a_0(x) \geq d > 0$, and let f be a function in $C^2(\bar{\Omega})$ satisfying

$$f > 0 \quad \text{in} \quad [0,\tilde{x}_0) \cup (\tilde{x}_1,1], \text{ for some } \tilde{x}_0,\tilde{x}_1 \in (0,1), \tilde{x}_0<\tilde{x}_1,$$
$$f < 0 \quad \text{in} \quad (\tilde{x}_0,\tilde{x}_1).$$

Then u satisfies

$$\begin{aligned} u &\leq 0 && \text{in } \Omega \\ L_\varepsilon u - f &= 0 && \text{in } (x_0(\varepsilon),x_1(\varepsilon)) \subset \Omega \\ u &= 0 && \text{in } \Omega\setminus(x_0(\varepsilon),x_1(\varepsilon)) , \end{aligned}$$

and the free boundary of u is given by

$$\begin{aligned} x_0(\varepsilon) - \tilde{x}_0 &= O(\sqrt{\varepsilon}) && \text{as } \varepsilon \to 0 \\ x_1(\varepsilon) - \tilde{x}_1 &= O(\sqrt{\varepsilon}) && \text{as } \varepsilon \to 0. \end{aligned}$$

Furthermore,

$$\|u-u_0\|_{L^2(\Omega)} \to 0 \qquad \text{as } \varepsilon \to 0,$$

where $u_0$ is given by (3.13).

## 4. Problems in $\mathbb{R}^1$. The case of non-symmetric bilinear forms.

We now consider the variational inequality (2.6) corresponding to the differential operator

$$L_\varepsilon = -\varepsilon\frac{d^2}{dx^2} + a_1(x)\frac{d}{dx} + a_0(x),$$

where $a_0, a_1 \in C^1(\bar{\Omega})$, $a_0(x) \geq 0$, $a_1(x) \geq d > 0$. We take $\Omega$ and f as in section 3. The bilinear form induced by $L_\varepsilon$

$$a(u,v) = \int_0^1 (\varepsilon\frac{du}{dx}\frac{dv}{dx} + a_1(x)\frac{du}{dx}v + a_0(x)uv)dx, \quad u,v \in H_0^1(\Omega),$$

is clearly non-symmetric.

Again by the results of section 2, the solution u of the variational inequality (2.6) belongs to $\mathfrak{F}$, so u satisfies:

$$\begin{aligned} u &\leq 0 \quad \text{in } \Omega \\ L_\varepsilon u - f &= 0 \quad \text{in some interval } (x_0, x_1) \supset (\tilde{x}_0, \tilde{x}_1) \\ u &= 0 \quad \text{in } \Omega\setminus(x_0, x_1). \end{aligned}$$

The arguments used in section 3 in order to obtain an asymptotic approximation of any $v \in \mathfrak{F}$, may be applied here to find that

(4.1) $$v_{as}(x) = v_0(x) - v_0(x_1)\exp(-a_1(x_1)\frac{x_1 - x}{\varepsilon}) + R(x_0, x_1, x, \varepsilon)$$
$$\text{if } x \in [x_0, x_1]$$

(4.2) $$v_{as}(x) = 0 \qquad \text{if } x \in [0, x_0) \cup (x_1, 1],$$

where

$$v_0(x) = \int_{x_0}^{x} \frac{f(s)}{a_1(s)} \exp(-\int_s^x \frac{a_0(t)}{a_1(t)}dt)ds$$

and

$$R(x_0, x_1, x, \varepsilon) = \frac{x_1 - x}{x_1 - x_0} v_0(x_1) \exp(-a_1(x_1)\frac{x_1 - x_0}{\varepsilon}).$$

Furthermore,

(4.3) $$v = v_{as} + O(\varepsilon),$$

where the order of magnitude estimate is in the supremum; cf. Eckhaus [4] for technical details.

We observe that the condition $v \leq 0$, as in section 3, immediately imposes a condition upon $x_0(\varepsilon)$. It is not difficult to show that

(4.4) $$x_0(\varepsilon) - \tilde{x}_0 = O(\varepsilon).$$

Thus, in the first approximation one part of the free boundary of u has been determined and proved to coincide within $O(\varepsilon)$ with the zero $\tilde{x}_0$ of f. However, contrary to the case of symmetric bilinear forms, the remaining part of the free boundary cannot be determined by elementary considerations alone. In what follows we shall show that the determination of $x_1(\varepsilon)$

can be achieved from a careful study of the functional J defined in section 2.

The following analysis of the elements of $\mathcal{F}$ will provide us with a clear picture of the behavior of the solution u of the variational inequality. We recall that each $v \in \mathcal{F}$ satisfies

(4.6) $v \leq 0$ in $\Omega$

(4.7) $L_\varepsilon v - f = 0$ in some interval $(x_0,x_1) \supset (\tilde{x}_0,\tilde{x}_1)$

(4.8) $v = 0$ in $\Omega \backslash (x_0,x_1)$.

Furthermore, to each $v \in \mathcal{F}$ there exists an asymptotic approximation (4.1), (4.2).

The following lemma is basic in our analysis. To get a concise formulation of the lemma we shall write $v(\cdot,x_0,x_1)$ for $v \in \mathcal{F}$ satisfying (4.6), (4.7), (4.8), and we shall denote the right-hand zero of $\tilde{v}_0(x) = \int_{\tilde{x}_0}^{x} \frac{f(s)}{a_1(s)} \exp(-\int_s^x \frac{a_0(t)}{a_1(t)}dt)ds$ by $\bar{x}_1$.

<u>Lemma 4.</u> Let $v_1(\cdot,x_0,x_1)$, $v_2(\cdot,x_0,x_2)$ be elements of $\mathcal{F}$ for fixed $x_1 < x_2 < \bar{x}_1$. Then $J(v_2) - J(v_1)$ will be negative, for $\varepsilon$ small enough, whenever $\tilde{v}_0(x_1)$ is negative.

<u>Proof.</u> Consider

(4.9) $J(v_2) - J(v_1) = -\frac{1}{2} \int_{x_0}^{x_1} (v_2-v_1)f\theta^{-2}dx - \frac{1}{2} \int_{x_1}^{x_2} v_2 f\theta^{-2}dx,$

where $\theta^{-2}(x) = \exp(-\frac{1}{\varepsilon} \int_0^x a_1(s)ds)$.

The contribution of the second integral is of order $O(\varepsilon\theta^{-2}(x_1))$. Now a very careful analysis of the first integral is needed. We denote the function $(v_2-v_1)\theta^{-2}$, defined on $[x_0,x_1]$, by w. Then (4.9) becomes

(4.10) $J(v_2) - J(v_1) = -\frac{1}{2} \int_{x_0}^{x_1} wf\theta^{-2}dx + O(\varepsilon\theta^{-2}(x_1))$.

Next, we apply the operator $L_\varepsilon$ to $w\theta^2$. This results in the following differential equation satisfied by w:

$$\varepsilon\frac{d^2w}{dx^2} + a_1(x)\frac{dw}{dx} - (a_0(x) - a_1'(x))w = 0.$$

Finally, set $w^*(x) = v_2^{-1}(x_1)\theta^2(x)w(x)$, then $w^*$ satisfies

(4.11) $\varepsilon\frac{d^2w^*}{dx^2} + a_1(x)\frac{dw^*}{dx} - (a_0(x) - a_1'(x))w^* = 0$ on $(x_0,x_1)$

(4.12) $w^*(x_0) = 0$ and $w^*(x_1) = 1$.

Note that for this problem the boundary layer occurs at $x_0$.

Now $w^*$ satisfies, in the supremum norm,

(4.13) $w^*(x) = w_0^*(x) - w_0^*(x_0)\exp(-a_1(x_0)\frac{x-x_0}{\varepsilon}) + O(\varepsilon)$,

where

(4.14) $w_0^*(x) = \frac{a_1(x_1)}{a_1(x)}\exp(-\int_x^{x_1}\frac{a_0(s)}{a_1(s)}ds)$

is a solution to the reduced equation, satisfying $w_0^*(x_1) = 1$.

We are now in a position to further analyse the functional J. Consider

$$J(v_2) - J(v_1) = -\tfrac{1}{2}\, v_2(x_1)\theta^{-2}(x_1)\int_{x_0}^{x_1} w_0^*(x)f(x)dx + O(\varepsilon\theta^{-2}(x_1)),$$

by a simple calculation one finds

$$J(v_2) - J(v_1) = -\tfrac{1}{2}\, v_2(x_1)\theta^{-2}(x_1)a_1(x_1)v_0(x_1) + O(\varepsilon\theta^{-2}(x_1)).$$

Moreover, for any $x \in [\tilde{x}_0, 1]$,

$$0 \leq v_0(x) - \tilde{v}_0(x) = \int_{x_0}^{\tilde{x}_0}\frac{f(s)}{a_1(s)}\exp(-\int_s^x \frac{a_0(t)}{a_1(t)}dt)ds = O(\varepsilon)$$

hence

$$J(v_2) - J(v_1) = -\tfrac{1}{2}\, v_2(x_1)\theta^{-2}(x_1)a_1(x_1)\tilde{v}_0(x_1) + O(\varepsilon\theta^{-2}(x_1)),$$

and the result follows, since $v_2(x_1) < 0, v_2 = O_s(1)$ and $\tilde{v}_0 = O_s(1)$ as $\varepsilon \to 0$.

Corollary 1. The right-hand part of the free boundary $x_1(\varepsilon)$ of u satisfies $\lim_{\varepsilon\downarrow 0}\min\{x_1(\varepsilon)-\bar{x}_1, 0\} = 0$, whenever $\bar{x}_1 \leq 1$. If $\bar{x}_1 > 1$, then $\lim_{\varepsilon\downarrow 0} x_1(\varepsilon) = 1$.

Corollary 2. The proof of the preceding lemma remains valid for any $f \in C^2(\bar{\Omega})$ satisfying

$$f > 0 \text{ in } [0, \tilde{x}_0), \qquad f = 0 \text{ in } [\tilde{x}_1, 1],$$
$$f < 0 \text{ in } (\tilde{x}_0, \tilde{x}_1).$$

In particular, the right-hand part of the free boundary of u satisfies $\lim_{\varepsilon\downarrow 0} x_1(\varepsilon) = 1$, since $\tilde{v}_0(x) < 0$ for any $x \in (\tilde{x}_0, 1]$.

The behavior of any $v \in \mathcal{S}$ satisfying $v(x_1) = 0$, $x_1 > \bar{x}_1$, can be treated by methods completely analogous to those of section 3. Here we find $\lim_{\varepsilon\downarrow 0}\max\{x_1(\varepsilon)-\bar{x}_1, 0\} = 0$ for each $v \in \mathcal{S}$.
Finally, having proved that $\lim_{\varepsilon\downarrow 0}|x_1(\varepsilon)-\bar{x}_1| = \delta(\varepsilon)$, where $\delta = o(1)$, for the right-hand free boundary $x_1(\varepsilon)$ of $u \in \mathcal{S}$, one may show that the assumption $\varepsilon = O(\delta(\varepsilon))$ leads to a contradiction.

We observe that if $\tilde{v}_0(x) < 0$ for any $x \in [\tilde{x}_0, 1]$ then the right-hand part of the free boundary occurs at $x = 1$.

Again we shall not pursue here the technical matter of determining higher approximations.

Finally, we define the function $\tilde{u}_0$ as follows

$$(4.15)\quad \tilde{u}_0(x)=\begin{cases}\tilde{v}_0(x) & \text{if } x \in [\tilde{x}_0,\bar{x}\,] \cap \bar{\Omega},\\ 0 & \text{if } x \in \bar{\Omega}\setminus[\tilde{x}_0,\bar{x}_1].\end{cases}$$

It is a matter of straightforward computation to show that

$$\|u-u_0\|_{L^2(\Omega)} \to 0 \text{ as } \varepsilon \to 0.$$

Summarizing, we have proved the following

Theorem 2. Let u be the solution of the variational inequality (2.6) induced by the differential operator $L_\varepsilon$ in $\Omega = (0,1)$, defined by

$$L_\varepsilon = -\varepsilon \frac{d^2}{dx^2} + a_1(x) \frac{d}{dx} + a_0(x),$$

where $a_0, a_1 \in C^1(\bar{\Omega}), a_0(x) \geqslant 0,\ a_1(x) \geqslant d > 0$ and f is an element of $C^2(\bar{\Omega})$ satisfying

$$f > 0 \quad \text{in } [0,\tilde{x}_0) \cup (\tilde{x}_1,1],\ \text{for some } \tilde{x}_0,\tilde{x}_1 \in (0,1), \tilde{x}_0<\tilde{x}_1,$$
$$f < 0 \quad \text{in } (\tilde{x}_0,\tilde{x}_1).$$

Then u satisfies

$$(4.16)\quad \begin{cases} u \leqslant 0 \text{ in } \Omega\\ L_\varepsilon u - f = 0 \text{ in } (x_0(\varepsilon),x_1(\varepsilon)) \subset \Omega\\ u = 0 \text{ in } \Omega\setminus(x_0(\varepsilon),x_1(\varepsilon))\end{cases}$$

and the free boundary of u is given by

$$(4.17)\quad x_0(\varepsilon) - \tilde{x}_0 = O(\varepsilon)$$
$$(4.18)\quad x_1(\varepsilon) - \bar{x}_1 = O(\varepsilon) \qquad \text{as } \varepsilon \to 0,$$

where $\bar{x}_1$ is the right-hand zero of the function $\tilde{v}_0$, given by

$$\tilde{v}_0(x) = \int_{\tilde{x}_0}^{x} \frac{f(s)}{a_1(s)} \exp\left(-\int_s^x \frac{a_0(t)}{a_1(t)}dt\right)ds,$$

whenever $\bar{x}_1 \leqslant 1$. If $\bar{x}_1 > 1$, then (4.18) is replaced by

$$(4.19)\quad x_1(\varepsilon) -1 = O(\varepsilon).$$

Furthermore,

$$\|u-u_0\|_{L^2(\Omega)} \to 0 \quad \text{as } \varepsilon \to 0,$$

where $\tilde{u}_0$ is given by (4.15).

References.

[1] Brézis, H.R. and G. Stampacchia, Sur la regularité de la solution d'inéquations elliptiques, Bull. Soc. Math. France 96 (1968), 153-180.

[2] Duvaut, G. and J.L. Lions, Inequalities in Mechanics and Physics, Springer-Verlag, Berlin 1976.

[3] Eckhaus, W., Boundary Layers in Linear Elliptic Singular Perturbation Problems, SIAM Review 14 (1972), 225-270.

[4] Eckhaus, W., Matched Asymptotic Expansions and Singular Perturbations. Math. Studies 6, North-Holland, Amsterdam, 1973.

[5] Eckhaus, W., Book on singular perturbations, to appear.

[6] Eckhaus, W. and E.M. de Jager, Asymptotic Solutions of Singular Perturbation Problems for Linear Differential Equations of Elliptic Type, Arch. Rational Mech. Anal. 23 (1966), 26-86.

[7] Lions, J.L. Partial Differential Inequalities, Russ. Math. Surveys 27 (1972), 91-159.

[8] Lions, J.L. Some Topics on Variational Inequalities and Applications, in: W. Eckhaus, ed., New Developments in Differential Equations, Mathematics Studies 21, North-Holland, Amsterdam 1976, 1-38.

[9] Lions, J.L. and E. Magenes, Non-homogeneous Boundary Value Problems and Applications, Springer-Verlag, Berlin 1972.

[10] Lions, J.L. and G. Stampacchia, Variational Inequalities, Comm. Pure Appl. Math. 20 (1967), 493-519.

[11] Protter, M.H. and H.F. Weinberger, Maximum Principles in Differential Equations, Prentice-Hall, Englewood Cliffs, 1967.

[12] Stampacchia, G. Formes bilinéaires coercitives sur les ensembles convexes, C.R. Acad. Sci. Paris 258 (1964), 4413-4416.

*Differential Equations and Applications*
*W. Eckhaus and E.M. de Jager (eds.)*


# HYPERBOLIC SINGULAR PERTURBATIONS OF NON LINEAR FIRST ORDER DIFFERENTIAL EQUATIONS

R. Geel (Ubbo Emmius Institute, Groningen)
and
E.M. de Jager (University of Amsterdam)

## 1. INTRODUCTION

We consider the non linear Cauchy problem:

$$(1.1)\quad \varepsilon\left(\frac{\partial^2 u}{\partial t^2} - \frac{\partial^2 u}{\partial x^2}\right) + a(x,t,u)\frac{\partial u}{\partial x} + b(x,t,u)\frac{\partial u}{\partial t} + d(x,t,u) = 0\ , \quad -\infty < x < +\infty\ ,\ t > 0\ .$$

$$(1.2)\quad u(x,0) = \frac{\partial u}{\partial t}(x,0) = 0\ .$$

$\varepsilon$ is a small positive parameter, $0 < \varepsilon << 1$.
The following conditions are imposed on the functions a,b en d.

i) $b > 0$ and $\frac{|a|}{b} < 1$ for all x, all non negative t, and all values of u.

ii) a, b and d are $C^\infty$ in (x,t) and $C^2$ in u.

REMARK

The regularity of a,b and d in (x,t) may be weakened; we suppose the strong regularity in order not to be obliged to keep a tedious bookkeeping in the subsequent analysis.

The so called reduced Cauchy problem reads:

$$(1.3)\quad a(x,t,w)\frac{\partial w}{\partial x} + b(x,t,w)\frac{\partial w}{\partial t} + d(x,t,w) = 0, -\infty < x < +\infty\ ,\ t > 0$$

$$(1.4)\quad w(x,0) = 0$$

Supposing that (1.3) - (1.4) has a solution which is "sufficiently" regular in a bounded closed domain $\Omega$, with part of the boundary along the x-axis, we put the key question in singular perturbation theory: to what extent is

$$(1.5)\quad \lim_{\varepsilon \downarrow 0} u = w\ ,$$

where the symbol "lim" has to be specified.
This question will be answered affirmatively: viz. there exists a solution u of (1.1) - (1.2) with the property that uniformly in $\Omega$

$$(1.6)\quad \begin{cases} u = w + O(\varepsilon) \\ \dfrac{\partial u}{\partial x} = \dfrac{\partial w}{\partial x} + O(\varepsilon) \\ \dfrac{\partial u}{\partial t} = \dfrac{\partial w}{\partial t} + \dfrac{\partial v}{\partial \tau} + O(\varepsilon) \end{cases}$$

where $v(x,\tau) = v(x,\frac{t}{\varepsilon})$ is a boundary layer term, concentrated in the upper neighbourhoud of $t = 0$.

Hyperbolic perturbations of non linear equations have also been studied recently by J.Genet and M.Madaune. These authors have considered initial boundary value problems for the equation

$$\varepsilon L_2[u] + L_1[u] + F(u) = f(t,x),\ (t,x) \in (0,T) \times \Omega, \Omega \subset\subset \mathbb{R}_n .$$

with

$$L_2 = \frac{\partial^2}{\partial t^2} - \Delta \quad \text{and} \quad L_1 = a(x,t)\frac{\partial}{\partial t} + \sum_{k=1}^{n} b_k(x,t)\frac{\partial}{\partial x_k} + c(x,t)$$

The non-linearity is confined to the non-linear function $F(u)$, e.g. $F(u) = |u|^{\rho}u$. We refer to lit. [1], [2] and [3].

## 2. A MODIFICATION OF A FIXED POINT THEOREM OF VAN HARTEN

In order to prove (1.6) we need the following modification of a fixed point theorem, given by van Harten [4], p.p. 189-192.
Let N be a normed linear space with norm $|\cdot|$, $u \in N$ and B a Banach space with norm $||\cdot||$, $v \in B$. Let F be a non linear mapping $N \to B$ with $F(o) = 0$ and with

$$F(u) = L(u) + \psi(u)$$

where L is the linearization of F at $u = 0$
The following conditions are imposed on L and $\psi$:

i) The mapping L from N to B is bijective and $L^{-1}$ is continuous

$$|L^{-1} v| \le \ell^{-1} ||v|| \quad , \forall v \in B.$$

ii) Let $\Omega_N(\rho) := \{ u \in N,\ |u| \le \rho \}$

$$\exists \bar{\rho} \text{ s.t. } ||\psi(u_1) - \psi(u_2)|| \le m(\rho)|u_1 - u_2|,\ \forall u_i \in \Omega_N(\rho),\ 0 \le \rho \le \bar{\rho},$$

where $m(\rho)$ is decreasing for $\rho$ decreasing with $\lim_{\rho \to o} m(\rho) = 0$

iii) $\rho_o := \sup_{\rho \ge 0} \{ \rho,\ 0 \le \rho \le \bar{\rho} \wedge m(\rho) \le \tfrac{1}{2}\ell \}$

ASSERTION

$f \in B \wedge ||f|| \le \tfrac{1}{2}\ell\rho_o$

$\exists\, u \in N$ with $F(u) = f$ and $|u| \le 2\ell^{-1} ||f||$

PROOF

The relation $F(u) = L(u) + \psi(u) = f$ is equivalent with the relation

$$v = Tv := f - \psi(L^{-1}v),$$

where $v = L(u)$
So the proof comes down to demonstration of the existence of a fixed point for the operator T.
Consider the ball $\Omega_B(\rho) = \{ v \in B,\ ||v|| \le \rho \}$. Whenever $||f|| \le \frac{1}{2}\ell\rho$, $0 \le \rho \le \rho_o$, T is srictly contractive in $\Omega_B(\ell\rho)$ and T maps $\Omega_B(\ell\rho)$

in $\Omega_B(\ell\rho)$ ; hence the existence of a fixed point v is guaranteed. For details of the proof see [4], pp. 189-192 or [5].

## 3. A PRIORI ESTIMATE FOR THE SOLUTION OF A LINEAR SINGULAR PERTURBATION PROBLEM OF HYPERBOLIC TYPE.

We consider the following linear Cauchy problem:

$$(3.1)\quad \varepsilon\left(\frac{\partial^2 u}{\partial t^2} - \frac{\partial^2 u}{\partial x^2}\right) + a(x,t;\varepsilon)\frac{\partial u}{\partial x} + b(x,t;\varepsilon)\frac{\partial u}{\partial t} + d(x,t;\varepsilon)\,u = f(x,t;\varepsilon)$$

$$-\infty < x < +\infty\ ,\ t > 0.$$

$$(3.2)\quad u(x,0) = g(x;\varepsilon)\ ,\ \frac{\partial u}{\partial t}(x,0) = h(x;\varepsilon).$$

$\varepsilon$ is again a small positive parameter, $0 < \varepsilon << 1$. We make the following assumptions:

i) For each compact subset $\Omega$ in $\{(x,t),\ t \geq 0\}$ there exists a positive constant $\varepsilon_1$, such that $a \in C^1(\Omega)$ , $b \in C^1(\Omega)$, $d \in C^0(\Omega)$ and $f \in C^0(\Omega)$, for every value of $\varepsilon$ with $0 < \varepsilon \leq \varepsilon_1$.

ii) $a$ , $\frac{\partial a}{\partial x}$ , $\frac{\partial a}{\partial t}$ , $b$ , $\frac{\partial b}{\partial x}$ , $\frac{\partial b}{\partial t}$ and $d$ are uniformly bounded for $(x,t,\varepsilon) \in \Omega \times (0,\varepsilon_1)$

iii) For each interval $[\alpha,\beta]$ there exists a positive constant $\varepsilon_2$ such that $g \in C^1[\alpha,\beta]$ and $h \in C^0[\alpha,\beta]$ for every value of $\varepsilon$ with $0 < \varepsilon \leq \varepsilon_2$ :

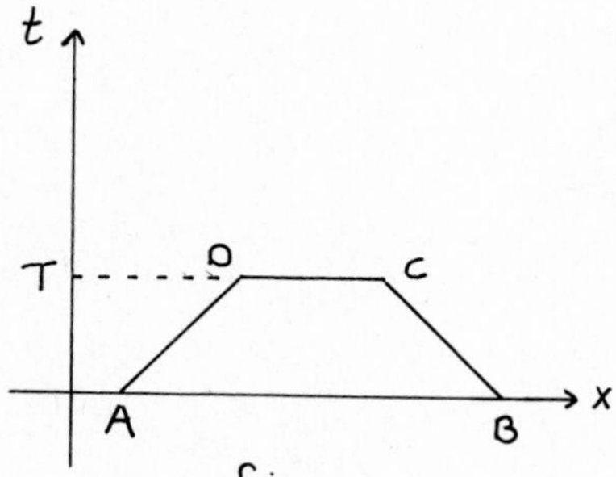

fig. 1

We take for $\Omega$ a domain as indicated in figure 1, with AD and BC characteristic. We have the following theorem, due to R.Geel.

Theorem

In any bounded closed domain $\Omega$ as indicated in figure 1, the solution u of the Cauchy problem (3.1) - (3.2) admits the following a priori estimate, valid for $\varepsilon$ sufficiently small, say $0 < \varepsilon \leq \varepsilon_0 \leq \min(\varepsilon_1,\varepsilon_2)$

$$(3.3)\quad \max_{\Omega}|u| + \varepsilon^{3/4}\max_{\Omega}\left|\frac{\partial u}{\partial x}\right| + \varepsilon^{3/4}\max_{\Omega}\left|\frac{\partial u}{\partial t}\right| \leq C(\Omega)\,\varepsilon^{-1/4}\,K(\Omega,\varepsilon)$$

where $C(\Omega)$ depends only on $\Omega$ and the coefficients a,b and d, but $C(\Omega)$ is independent of $\varepsilon$, and where $K(\Omega,\varepsilon)$ is given by:

$$(3.4)\quad K(\Omega,\varepsilon) = \max_{\Omega}|f| + \max_{AB}|g| + \varepsilon^{1/2}\max_{AB}|g_x| + \varepsilon^{1/2}\max_{AB}|h|$$

PROOF

The proof is established by using energy integrals (cf. [6],[7] ; for a detailed proof see [8] or [5]

REMARK

By this theorem is clear that a formal approximation $\tilde{u}$ of a solution of (3.1) and (3.2) - i.e. a function $\tilde{u}$ which satisfies (3.1) - (3.2) up to $O(\varepsilon^2)$, uniformly in some closed domain $\Omega$ (see fig. 1) - approximates the solution u ; in fact we have uniformly in $\Omega$,

$$\left\{\begin{aligned} u &= \tilde{u} + O(\varepsilon) \\ \frac{\partial u}{\partial x} &= \frac{\partial \tilde{u}}{\partial x} + O(\varepsilon) \\ \frac{\partial u}{\partial t} &= \frac{\partial \tilde{u}}{\partial t} + O(\varepsilon) \end{aligned}\right. \tag{3.5}$$

(See ref.[8] or [5]).

## 4. FORMAL APPROXIMATION OF THE CAUCHY PROBLEM.

We consider the Cauchy problem of section 1.

$$L_\varepsilon(u) = \varepsilon\left(\frac{\partial^2 u}{\partial t^2} - \frac{\partial^2 u}{\partial x^2}\right) + a(x,t,u)\frac{\partial u}{\partial x} + b(x,t,u)\frac{\partial u}{\partial t} + d(x,t,u) = 0\ , \quad t > 0 \tag{4.1}$$

$$u(x,0) = 0 \tag{4.2}$$

$$\frac{\partial u}{\partial t}(x,0) = 0 \tag{4.3}$$

Suppose $w_1(x,t)$ is a solution of the reduced problem (1.3) - (1.4). $w_1(x,t)$ does not satisfy the second initial condition. In order to fullfill this condition we add to $w_1$ a correction term $\varepsilon v_1(x,\tau) = \varepsilon v_1(x,\frac{t}{\varepsilon})$, where $v_1$ satisfies the linear boundary value problem:

$$\frac{\partial^2 v_1}{\partial \tau^2} + b(x,0,w_1(x,0))\frac{\partial v_1}{\partial \tau} = 0 \tag{4.4}$$

$$\varepsilon\frac{\partial v_1}{\partial t}(x,0) = \frac{\partial v_1}{\partial \tau}(x,0) = -\frac{\partial w_1}{\partial t}(x,0) \tag{4.5}$$

$$\lim_{\tau\to\infty} v_1(x,\tau) = 0. \tag{4.6}$$

$$\text{Hence } v_1(x,\tau) = \frac{\frac{\partial w_1}{\partial t}(x,0)}{b(x,0,w_1(x,0))} \exp - \left[b(x,0,w_1(x,0))\frac{t}{\varepsilon}\right], \tag{4.7}$$

and $v_1$ is of boundary layer type.
It is now readily verified that

$$\tilde{u}_1 = w_1 + \varepsilon v_1 \tag{4.8}$$

is a formal approximation of u ; in fact, we have

$$L_\varepsilon(\tilde{u}_1) = O(\varepsilon)\ ,\ \tilde{u}_1(x,o) = \varepsilon\, v_1(x,0)\ ,\ \frac{\partial \tilde{u}_1}{\partial t}(x,0) = 0\ ,$$

uniformly valid in any bounded domain, with part of the boundary along the x-axis, under the assumption that $w_1$ is sufficiently regular in this domain (we need $w \in C^3(x,t)$)

A sharper formal approximation:

$$\tilde{u}_2 = w_1(x,t) + \varepsilon\, w_2(x,t) + \varepsilon\, v_1(x,\tau) + \varepsilon^2\, v_2(x,\tau) \tag{4.9}$$

can be obtained similarly after expanding a,b and d into powers of $(u-w_1)$ and equating equal powers of $\varepsilon$.

$w_2$ satisfies the linear initial value problem:

$$a(x,t,w_1)\frac{\partial w_2}{\partial x} + b(x,t,w_1)\frac{\partial w_2}{\partial t} + \tilde{d}(x,t,w_1)\, w_2 = -\frac{\partial^2 w_1}{\partial t^2} - \frac{\partial^2 w_1}{\partial x^2}\ ,\ t > 0$$

$$w_2(x,0) = -v_1(x,0),$$

where

$$\tilde{d}(x,t,w_1) := \frac{\partial a}{\partial u}(x,t,w_1)\frac{\partial w_1}{\partial x} + \frac{\partial b}{\partial u}(x,t,w_1)\frac{\partial w_1}{\partial t} + \frac{\partial d}{\partial u}(x,t,w_1)$$

The correction term $\varepsilon^2 v_2(x,\tau)$ satisfies the linear boundary value problem

$$\frac{\partial^2 v_2}{\partial \tau^2} + b(x,0,w_1(x,0))\frac{\partial v_2}{\partial \tau} = -\Phi(x,\tau)$$

$$\frac{\partial v_2}{\partial \tau}(x,0) = -\frac{\partial w_2}{\partial t}(x,0)\ ,\ \lim_{\tau\to\infty} v_2(x,\tau) = 0\ ,$$

where

$$\begin{aligned}\Phi(x,\tau) := {} & [\frac{\partial a}{\partial u}(x,0,w_1(x,0))\frac{\partial w_1}{\partial x}(x,0) + \frac{\partial b}{\partial u}(x,0,w_1(x,0))\frac{\partial w_1}{\partial t}(x,0)\\ & + \frac{\partial d}{\partial u}(x,0,w_1(x,0))]\, v_1(x,\tau) + \\ & + [\{\frac{\partial b}{\partial t}(x,0,w_1(x,0)) + \frac{\partial b}{\partial u}(x,0,w_1(x,0))\frac{\partial w_1}{\partial t}(x,0)\}\,\tau + \\ & + \frac{\partial b}{\partial u}(x,0,w_1(x,0))\, w_2(x,0)]\frac{\partial v_1}{d\tau} + \\ & + a(x,0,w_1(x,0))\frac{\partial v_1}{\partial x} + \frac{\partial b}{\partial u}(x,0,w_1(x,0))\, v_1\frac{\partial v_1}{\partial \tau}\end{aligned}$$

From these equations it follows immediately that also $v_2(x,\frac{t}{\varepsilon})$ is of boundary layer type, concentrated in an upper neighbourhood of the x-axis.

In case $w_1$ is sufficiently regular in a bounded domain D with part of the boundary along the x-axis, then also $w_2, v_1$ and $v_2$ are sufficiently regular such that

$$(4.10)\quad \begin{cases} L_\varepsilon(\tilde{u}_2) = O(\varepsilon^2) \\ \tilde{u}_2(x,0) = \varepsilon^2\, v_2(x,0) \\ \frac{\partial u2}{\partial t}(x,0) = 0 \quad , \end{cases}$$

uniformly in D.
In order not to be forced to make many tedious calculations we assume $w_1(x,t) \in C^\infty$ in the domain of consideration D. Given the differential equation, this assumption restricts in general the size and the location of D.
We put now

$$(4.11)\quad u = \tilde{u}_2 + R \quad ,$$

and we shall show by means of van Harten's theorem of section 2 that $R = O(\varepsilon^{7/4})$ uniformly in any bounded closed domain $\Omega$ of the type of fig.1, where $w_1$ is a $C^\infty$ function.

## 5. ESTIMATION OF THE REMAINDER TERM R.

Due to (4.11) we have $L_\varepsilon(\tilde{u}_2+R) = 0$ or due to (4.10)

$$(5.11)\quad \begin{aligned} F_{\varepsilon,\tilde{u}}(R) := {} & \varepsilon\left(\frac{\partial^2 R}{\partial t^2} - \frac{\partial^2 R}{\partial x^2}\right) + a(\tilde{u} + R)\frac{\partial R}{\partial x} + b(\tilde{u} + R)\frac{\partial R}{\partial t} \\ & + [\{a(\tilde{u} + R) - a(\tilde{u})\}\frac{\partial \tilde{u}}{\partial x} + \{\, b(\tilde{u} + R) - b(\tilde{u})\}\frac{\partial \tilde{u}}{\partial t} \\ & + d(\tilde{u} + R) - d(\tilde{u})] = O(\varepsilon^2) \ , \text{ uniformly in } \Omega \text{ (fig.1)} \end{aligned}$$

and

$$R(x,0) = -\,\varepsilon^2\, v_2(x,0) \ , \ \frac{\partial R}{\partial t}(x,0) = 0$$

We have omitted the subscript of $\tilde{u}$ and the arguments x and t in the coefficients a,b and d.
Putting

$$(5.2)\quad R^* = R + \varepsilon^2\, v_2(x,0) = R + O(\varepsilon^2), \text{ and}$$
$$\tilde{u}^* + R^* = \tilde{u} + R \text{ or } \tilde{u}^* = \tilde{u} + O(\varepsilon^2)$$

we obtain for $R^*$ the differential equation,

$$(5.3)\quad F_{\varepsilon,\tilde{u}^*}(R^*) = O(\varepsilon^2) \ , \text{ uniformly in } \Omega$$

with the initial conditions

$$(15.4)\quad R^*(x,0) = \frac{\partial R^*}{\partial t}(x,0) = 0 \ .$$

In order to get the right estimate for R within the order $O(\varepsilon^2)$ we may estimate as well $R^*$ instead of R.
We now apply the theorem of section 2.

We take

(5.5) $N = \{u; u \in C^1(\Omega), \ \frac{\partial^2 u}{\partial t^2} - \frac{\partial^2 u}{\partial x^2} \in C^0(\Omega), \ u(x,0) = \frac{\partial u}{\partial t}(x,0) = 0\}$,

where weak second derivatives are allowed, with

$$|u|_\varepsilon = \max_\Omega |u| + \varepsilon^{3/4} \max_\Omega |u_x| + \varepsilon^{3/4} \max_\Omega |u_t|,$$

and

(5.6) $B = \{ v; v \in C^0(\Omega)\}$ with $||v|| = \max_\Omega |v|$

(In the right hand sides of the norm definitions absolute values have been taken)

$F_{\varepsilon,\tilde{u}^*}$ is a non linear map from N to B with $F_{\varepsilon,\tilde{u}^*}(0) = 0$. The conditions i), ii) and iii) of the theorem of section 2 are fulfilled for the operator $F_{\varepsilon,\tilde{u}^*}$

Ad i) $L : N \to B$ is clearly bijective and $L^{-1}$ is continuous according to the theory of linear hyperbolic differential equations. Using the theorem of section 3 we have for $\varepsilon$ sufficiently small

$$| L^{-1} v |_\varepsilon \le C(\Omega)\, \varepsilon^{-\frac{1}{4}} ||v||$$

and hence

(5.7) $\ell^{-1} = C\, \varepsilon^{-\frac{1}{4}}$, with C a generic constant independent of $\varepsilon$.

Ad ii) $\psi(u)$ satisfies the Lipschitz condition and a small calculation yields

$m(\rho) = C \cdot \rho\, \varepsilon^{-3/4}$, with C again a generic constant, independent of $\varepsilon$.

Ad iii) $m(\rho_o) = \frac{1}{2}\ell$ or

(5.8) $\rho_o = \frac{1}{2C^2}\varepsilon$

In order to estimate $R^*$ we remark that

(5.9) $F_{\varepsilon,\tilde{u}^*}(R^*) = f < C_1\varepsilon^2$, uniformly in $\Omega$ for $\varepsilon$ sufficiently small; $C_1$ independent of $\varepsilon$.

Hence for $\varepsilon$ sufficiently small, say $0 < \varepsilon \le \varepsilon_o$

(5.10) $||f|| \le \frac{1}{2}\ell\rho_o = \frac{1}{4C^3}\varepsilon^{5/4}$

Application of van Harten's theorem gives finally

$$|R^*|_\varepsilon \le 2\,\ell^{-1}\,||f|| = O(\varepsilon^{7/4}), \text{ uniformly in } \Omega$$

or due to (5.2)

(5.11) $|R|_\varepsilon = O(\varepsilon^{7/4})$, uniformly in $\Omega$.

## 6. Conclusions

From sections 4 and 5 it follows now that in any bounded closed domain $\Omega$ of the type of fig. 1, where $w_1$ is $C^\infty$, there exists a solution of the initial value problem (1.1) - (1.2) with the properties:

$$u(x,t) = w_1(x,t) + \varepsilon w_2(x,t) + \varepsilon\, v_1(x,\tfrac{t}{\varepsilon}) + \varepsilon^2\, v_2(x,\tfrac{t}{\varepsilon}) + O(\varepsilon^{7/4})$$

$$= w_1(x,t) + O(\varepsilon)$$

$$\frac{\partial u}{\partial x}(x,t) = \frac{\partial w1}{\partial x}(x,t) + \varepsilon\, \frac{\partial w2}{\partial x}(x,t) + \varepsilon\, \frac{\partial v1}{\partial x}(x,\tfrac{t}{\varepsilon}) + \varepsilon^2\, \frac{\partial v2}{\partial x}(x,\tfrac{t}{\varepsilon}) + O(\varepsilon)$$

$$= \frac{\partial w1}{\partial x}(x,t) + O(\varepsilon)$$

$$\frac{\partial u}{\partial t}(x,t) = \frac{\partial w_1}{\partial t}(x,t) + \varepsilon\, \frac{\partial w_2}{\partial t}(x,t) + \frac{\partial v_1}{\partial \tau}(x,\tau) + \varepsilon\, \frac{\partial v_2}{\partial \tau}(x,\tau) + O(\varepsilon)$$

$$= \frac{\partial w_1}{\partial t}(x,t) + \frac{\partial v_1}{\partial \tau}(x,\tau) + O(\varepsilon),$$

uniformly valid in $\Omega$.

## LITERATURE


1. J. Genet, M. Madaune — Perturbations singulières pour une classe de problèmes hyperboliques non linéaires, C.R. Ac. Sc., Paris, 283 (1976) A - 167 - 170.

2. J. Genet, M. Madaune — Perturbations singulières pour une classe de problèmes hyperboliques non linéaires, Singular Perturbations and Boundary Layer Theory, Lyon 1976, Lecture Notes in Mathematics 594, Springer.

3. M. Madaune — Perturbations singulières de type hyperbolique - hyperbolique non linéaire. Publications Mathematiques de Pau, 1977.

4. A. van Harten — Singularly perturbed non-linear 2nd order elliptie boundary value problems. Thesis, University of Utrecht, 1975.

5. R. Geel — Singular Perturbations of hyperbolic type, Thesis, University of Amsterdam, 1978.

6. L. Bers, F. John, M. Schechter — Partial Differential Equations, Lectures in Applied Mathematics, Interscience Publ., 1964.

7. E.M. de Jager Singular Perturbations of hyperbolic type. Nieuw Archief voor Wiskunde (3), Vol. XXIII, 1975.

8. R. Geel Singular Perturbations of hyperbolic type. Report 76-19 , University of Amsterdam, 1976.

R. Geel
Instituut voor Lerarenopleiding
Ubbo Emmius
Paddepoel
Groningen (the Netherlands)

E.M. de Jager
Mathematical Institute
University of Amsterdam
Roetersstraat 15
Amsterdam (the Netherlands)

*Differential Equations and Applications*
*W. Eckhaus and E.M. de Jager (eds.)*


# THE NATURE OF THE "ACKERBERG - O'MALLEY RESONANCE"

P.P.N. de Groen
Eindhoven University of Technology
Department of Mathematics
Eindhoven, The Netherlands

## 1. INTRODUCTION

In this paper we give a survey of the results we have obtained in [1] on the asymptotics of the singularly perturbed two-point second-order boundary value problem of turning type:

(1.1.a) $L_\varepsilon u := -\varepsilon u'' + xp(x,\varepsilon)u' + q(x,\varepsilon)u = 0$ , $(' = d/dx)$

(1.1.b) $u(a) = A$ , $u(b) = B$ , $0 < \varepsilon << 1$ ,

where $a < 0 < b$ and where p, 1/p and q are smooth real functions on the set $[a,b] \times (0, \varepsilon_0]$. For normalization we assume p positive and $p(0,0) = 1$. The case p negative can be treated analogously. The problem is to construct an asymptotic approximation to the solution of (1.1) and to prove its validity.

Ackerberg & O'Malley have drawn attention to this problem in [2]; they construct a formal approximation by the WKBJ method. They remark that this approximation exhibits "resonance", i.e. that this approximation does not decay exponentially fast for $\varepsilon \to +0$ , if $q(0,0)/p(0,0)$ is equal to a non-positive integer. Most publications that followed this one fail in giving a better insight in this phenomenon since they overlook the important question of existence and uniqueness of the solution which they try to approximate. The first satisfactory explanation of resonance is given in [3], namely that $L_\varepsilon$ has an eigenvalue $\lambda(\varepsilon)$ which converges to zero for $\varepsilon \to +0$ if and only if $q(0,0)$ is equal to a nonpositive integer. Hence problem (1.1) has a unique solution if and only if $\lambda(\varepsilon) \neq 0$; if $|\lambda(\varepsilon)|$ is small enough, this solution exhibits resonance, i.e. the nearby free mode dominates the solution. Here we shall construct a uniformly valid approximation to the solution of (1.1), provided $\lambda(\varepsilon)\neq 0 \quad \forall \varepsilon \in (0,\varepsilon_0]$, and we shall give an idea of the techniques we have used. For detailed proofs we refer to [1]. We remark that our results can easily be generalized to problems, where p is negative, to problems with several turning points and to analogous elliptic problems in several dimensions.

## 2. THE RESULTS

The operator $L_\varepsilon$ of equation (1.1.a) is selfadjoint with respect to the weighted inner product

$$(2.1) \qquad (u,v)_\varepsilon := \int_a^b u(x)\bar{v}(x)w_\varepsilon(x)dx , \qquad w_\varepsilon(x) := \exp\{-\frac{1}{\varepsilon}\int_0^x tp(t,\varepsilon)dt\} .$$

The well-known Sturm-Licuville theory implies that $L_\varepsilon$ has a countable set of simple eigenvalues ,

$$\{\lambda_k(\varepsilon) \mid k = 0,1,2,\ldots\} \qquad (\text{assume } \lambda_{k+1} > \lambda_k \ \forall k) ,$$

and that the corresponding set of orthogonal eigenfunctions,

$$\{e_k(x,\varepsilon) \mid k = 0,1,2,\ldots\} ,$$

which satisfy

$$L_\varepsilon e_k = \lambda_k e_k , \quad e_k(a,\varepsilon) = e_k(b,\varepsilon) = 0 , \quad (e_k,e_j)_\varepsilon = s_k^2\delta_{kj} ,$$

is complete in the set of square integrable functions on (a,b). Approximations of $\lambda_k$ and $e_k$ can be found by solving the equation $L_\varepsilon u = \lambda u$ approximately. Because the main weight of $w_\varepsilon$ is located in an $\mathcal{O}(\sqrt{\varepsilon})$-neighbourhood of $x = 0$, we apply the stretching $\xi := x/\sqrt{2\varepsilon}$ and we find Hermite's equation as the highest order part of the stretched equation:

$$(2.2) \qquad -\ddot{u} + 2\xi\dot{u} - 2(\lambda + q(0,0))u = 0 , \qquad (\dot{} = d/d\xi) .$$

If $\lambda$-q(0,0) is a nonnegative integer, say k, one solution of (2.2) is of polynomial growth for $\xi \to \pm\infty$ , namely the k-th Hermite polynomial $H_k$; otherwise all solutions are of exponential growth, namely of the order $\mathcal{O}(\exp \xi^2)$ for $\xi \to \pm\infty$ . Since it is impossible to match these exponentially growing solutions to the boundary values u(a) = u(b) = 0 in boundary layers near x = a and x = b, we can construct formal approximations to eigenvalues and eigenfunctions only if $\lambda$ - q(0,0) is integral. This leads to the result:

THEOREM 1 : *The eigenvalues and eigenfunctions satisfy for each* k

$$(2.2.a) \qquad \lambda_k(\varepsilon) = k + q(0,0) + \mathcal{O}(\varepsilon) , \qquad \varepsilon \to +0$$

$$(2.2.b) \qquad \| s_k^{-1}e_k(x,\varepsilon) - (2\pi\varepsilon)^{-\frac{1}{4}}(2^n n!)^{-\frac{1}{2}}H_k(x/\sqrt{2\varepsilon}) \|_\varepsilon = \mathcal{O}(\sqrt{\varepsilon}) , \qquad \varepsilon \to +0$$

*where* $H_k$ *is the* k-*th Hermite polynomial and where* $\|\cdot\|_\varepsilon$ *denotes the norm corresponding to the inner product* $(\cdot,\cdot)_\varepsilon$ .

$H_k(x/\sqrt{2\varepsilon})$ needs not be a good approximation of $e_k$ outside an $\mathcal{O}(\sqrt{\varepsilon})$-neighbourhood of $x = 0$. We point out that in estimate (2.2.b) this is not necessary, since the weight function $w_\varepsilon$ in the norm $\|\cdot\|_\varepsilon$ is exponentially small outside an $\mathcal{O}(\sqrt{\varepsilon})$-neighbourhood of $x = 0$. A better approximation is obtained by solving the reduced equation

$$xpu' + qu = \lambda_k(0)u \tag{2.3}$$

for $x \neq 0$ and by matching it to the approximation $H_k(x/\sqrt{2\varepsilon})$ in a neighbourhood of $x = 0$ and to the boundary values $u(a) = u(b) = 0$ at $x = a$ and $x = b$. The solution of (2.3) is

$$v_k(x) := c_k x^k \exp\{\int_0^x (k - kp(t,0) - q(t,0) + q(0,0)) \frac{dt}{tp(t,0)}\}.$$

Choosing the scaling factor $s_k$ (the length of $e_k$) by

$$s_k := (2\pi\varepsilon)^{\frac{1}{4}}(2^n n!)^{\frac{1}{2}}$$

we find from matching to the approximation $H_k$ in a neighbourhood of $x = 0$ the integration constant

$$c_k := (2/\varepsilon)^{\frac{1}{2}k}.$$

Matching of $v_k$ to the boundary conditions at $x = a$ and $x = b$ yields the boundary layer approximations (with $\theta = a$ or $b$)

$$z_\varepsilon(x,\theta) = -v_k(\theta)\exp\{\theta p(\theta,0)\frac{x-\theta}{\varepsilon}\}.$$

Adding the constructed approximations and subtracting the common term $c_k x^k$ of $H_k$ and $v_k$ we find a uniformly valid approximation:

THEOREM 2: *For every* $k$ *a constant* $C$ *exists, such that*

$$|e_k(x,\varepsilon) - H_n(x/\sqrt{2\varepsilon}) + c_k x^k - v_k(x) - z_\varepsilon(x,a) - z_\varepsilon(x,b)| \leq C\varepsilon^{\frac{1}{2}}(1 + \varepsilon^{-\frac{1}{2}n}x^n) \tag{2.4}$$

*for all* $x \in [a,b]$ *and* $\varepsilon \in (0,\varepsilon_0]$ *uniformly.*

Now we are in position to construct a uniformly valid approximation to the solution $U_\varepsilon$ of problem (1.1). We can construct formal approximations of the solution $U_\varepsilon$ of (1.1), which consist of boundary layer terms at $x = a$

and x = b only and which are exponentially small in the interior. The lowest order part is $R_\varepsilon$,

(2.5) $$R_\varepsilon(x) := A\rho(ax)\exp\{ap(a,0)(x-a)/\varepsilon\} + B\rho(bx)\exp\{bp(b,0)(x-b)/\varepsilon\},$$

where $\rho$ is a $C^\infty$ cut-off function satisfying $\rho(x) = 1$ if $x > \frac{1}{2}$ and $\rho(x) = 0$ if $x < 0$. The fact that $R_\varepsilon$ is a formal approximation of $U_\varepsilon$ means that $L_\varepsilon(U_\varepsilon - R_\varepsilon)$ is small. If all eigenvalues of $L_\varepsilon$ are well-separated from zero, i.e. if $q(0,0)$ is not a nonpositive integer, then this smallness implies that $U_\varepsilon - R_\varepsilon$ itself is small. If $q(0,0) = -k$ is a nonpositive integer, then $\lambda_k \to 0$ for $\varepsilon \to +0$, while all other eigenvalues remain well-separated from zero. In such a case only the component of $U_\varepsilon - R_\varepsilon$, orthogonal to $e_k$, is small; the projection onto $e_k$ satisfies

$$s_k^{-2}(U_\varepsilon - R_\varepsilon, e_k)_\varepsilon e_k = s_k^{-2}\lambda_k^{-1}(U_\varepsilon - R_\varepsilon, L_\varepsilon e_k)_\varepsilon e_k =$$
$$= s_k^{-2}\lambda_k^{-1}(L_\varepsilon(U_\varepsilon - R_\varepsilon), e_k)_\varepsilon e_k ,$$

provided $\lambda_k(\varepsilon) \neq 0$. Approximating the inner product with the aid of the approximations constructed before we find the result:

THEOREM 3: *Let* k *be the nonnegative integer that is nearest to* $q(0,0)$. *If* $\lambda_k(\varepsilon) \neq 0$ *problem* (1.1) *has a unique solution* $U_\varepsilon$ *which satisfies for* $\varepsilon \to +0$

(2.6) $$U_\varepsilon(x) = R_\varepsilon(x) + \frac{e_k(x)}{\lambda_k(\varepsilon)}\{Bw_\varepsilon(b)v_k(b) + Aw_\varepsilon(a)v_k(a)\}(1 + \mathcal{O}(\sqrt{\varepsilon})) +$$

$$+ \begin{cases} \mathcal{O}(\varepsilon A w_\varepsilon^{\frac{1}{2}}(a)) + \mathcal{O}(\varepsilon B w_\varepsilon^{\frac{1}{2}}(b) w_\varepsilon^{-\frac{1}{2}}(x)) & \text{if } x \geq 0 \\ \mathcal{O}(\varepsilon B w_\varepsilon^{\frac{1}{2}}(b)) + \mathcal{O}(\varepsilon A w_\varepsilon^{\frac{1}{2}}(a) w_\varepsilon^{-\frac{1}{2}}(x)) & \text{if } x \leq 0 \end{cases}$$

*uniformly with respect to* $x \in [a,b]$; $w_\varepsilon$ *is the weight function defined in* (2.1).

This theorem displays precisely how the mechanism of "resonance" works. If $\lambda_k(\varepsilon)$ is far from zero, the eigenfunction $e_k$ is multiplied by an exponentially small term, such that its contribution is small with respect to $R_\varepsilon$. Only if $\lambda_k(\varepsilon)$ is of the order $\mathcal{O}(w_\varepsilon(a) + w_\varepsilon(b))$, the resonant part dominates the solution.

## 3. THE TECHNIQUES

In order to get an idea of the techniques by which the theorems 1, 2 and 3 can be proved, we consider the special problem

$$(3.1) \qquad \Pi_\varepsilon u := -\varepsilon u'' + xu' + qu = 0 , \quad u(-1) = A , \quad u(1) = B ,$$

where q is a constant. $\Pi_\varepsilon$ is selfadjoint with respect to the inner product (2.1) if $w_\varepsilon(x) := \exp(-\tfrac{1}{2}x^2/\varepsilon)$. We shall denote again its eigenvalues and eigenfunctions by $\lambda_k$ and $e_k$. Formula (2.4) yields the formal approximation $\chi_k$ of $e_k$ in this case,

$$\chi_k(x,\varepsilon) := H_k(x/\sqrt{2\varepsilon}) - H_k(1/\sqrt{2\varepsilon})\{\rho(x)\exp(\frac{x-1}{\varepsilon}) + (-1)^k\rho(-x)\exp(\frac{-1-x}{\varepsilon})\},$$

where $\rho$ is a cut-off function as in (2.5). It is easily seen that this function satisfies

$$(3.2) \qquad (\chi_k,\chi_m)_\varepsilon = (2\pi\varepsilon)^{\frac{1}{2}}2^k k!\{\delta_{km} + \mathcal{O}(\varepsilon^{-\frac{1}{2}k-\frac{1}{2}m}\exp(-1/2\varepsilon))\} ,$$

$$(3.3) \qquad \|(\Pi_\varepsilon - q - k)\chi_k\|_\varepsilon^2 = \mathcal{O}(\varepsilon^{1-n}\exp(-1/2\varepsilon)) ,$$

$$(3.4) \qquad ((\Pi_\varepsilon - q - k)\chi_k,\chi_k)_\varepsilon = 2^{k+1}\varepsilon^{-k}\exp(-1/2\varepsilon)(1 + \mathcal{O}(\varepsilon)) .$$

The eigenvalues of a selfadjoint operator are characterized by the minimax and maximin properties of Rayleigh's quotient, cf. [4]. Let $\mathcal{D}$ denote the set of all complex-valued functions on [-1,1], which satisfy $u(\pm 1) = 0$, and whose second derivative is square integrable. The k-th eigenvalue satisfies:

$$(3.5) \qquad \lambda_k(\varepsilon) = \inf_{E\subset\mathcal{D},\dim E\geq k+1} \ \sup_{u\in E,u\neq 0} (\Pi_\varepsilon u,u)_\varepsilon/\|u\|_\varepsilon^2 ,$$

$$(3.6) \qquad \lambda_k(\varepsilon) = \sup_{F\subset L^2(-1,1),\dim F\leq k} \ \inf_{u\in F^\perp\cap\mathcal{D},u\neq 0} (\Pi_\varepsilon u,u)_\varepsilon/\|u\|_\varepsilon^2 .$$

The supremum and infimum are attained if E and F are the spans of the first k + 1 and k eigenfunctions respectively. Any other set E or F yields an upper or lower estimate for $\lambda_k$ respectively. So, in conjunction with (3.2) and (3.4), the minimax criterion implies:

$$(3.7) \qquad \lambda_k(\varepsilon) \leq \sup_{u\in\operatorname{span}\{\chi_0,\dots,\chi_k\},\|u\|_\varepsilon=1} (\Pi_\varepsilon u,u)_\varepsilon \leq k + q + C\varepsilon^{-n-\frac{1}{2}}e^{-1/2\varepsilon}$$

where C is independent of $\varepsilon$.
For a lower bound we stretch the x-variable in the eigenvalue equation $\Pi_\varepsilon u = \lambda u$ by taking $\xi := x/\sqrt{2\varepsilon}$; we obtain

$$\tilde{\Pi}_\varepsilon v := -\ddot{v} + 2\xi\dot{v} + 2qv = 2\lambda v\ , \qquad v(\pm 1/\sqrt{2\varepsilon}) = 0 \qquad (^\cdot = d/d\xi)\ .$$

It is easily seen that the Rayleigh quotient of this operator $\tilde{\Pi}_\varepsilon$ cannot increase, and hence that $\lambda_k(\varepsilon)$ cannot increase, if $\varepsilon$ decreases. In the limit for $\varepsilon \to +0$ the operator $\tilde{\Pi}_\varepsilon$ tends to the well-known Hermite operator (harmonic oscillator), whose eigenvalues are the nonnegative integers (plus q) and whose eigenfunctions are the Hermite polynomials, which are complete. This implies $\lambda_k \geq k + q$; in conjunction with (3.7) this proves (2.2.a) for this special case. In the general problem an upper bound is obtained in the same way as in (3.7). A lower bound is obtained by application of the maximin criterion to the orthogonal complement of $\{\chi_0,\ldots,\chi_{k-1}\}$ and by comparison of the Rayleigh quotients of $L_\varepsilon$ and $\Pi_\varepsilon$.

Let us now assume that $\lambda_k(\varepsilon)$ is the eigenvalue that is nearest to zero, then (2.2.a) tells us that the distance from all other eigenvalues to zero is larger than ½. Expanding the difference between the true and the approximate eigenfunction into the eigenfunctions of $\Pi_\varepsilon$ we find by (3.3):

$$\|\chi_k - s_k^{-2}(\chi_k,e_k)e_k\|_\varepsilon^2 = \sum_{j=0,j\neq k}^{\infty} |(\chi_j,e_j/s_j)_\varepsilon|^2 =$$

$$= \sum_{j=0,j\neq k}^{\infty} \left|\frac{((\Pi_\varepsilon - k - q)\chi_k,e_j)_\varepsilon}{s_j(\lambda_j - k - q)}\right|^2 \leq 4\|(\Pi_\varepsilon - k - q)\chi_k\|_\varepsilon^2 =$$

$$= \mathcal{O}(\varepsilon^{1-n}\exp(-1/2\varepsilon))\ ,$$

which proves the validity of the formal approximation in $\|\cdot\|_\varepsilon$-norm. For a proof of validity of (2.6) in $\|\cdot\|_\varepsilon$-norm we operate in the same way.

Pointwise error estimates for the approximations of the eigenfunctions of $\Pi_\varepsilon$ and of solutions of (3.1) can be obtained using Sobolev's inequality and the maximum principle. Integrating by parts we find for each $u \in \mathcal{D}$:

$$(\Pi_\varepsilon u,u)_\varepsilon = \int_{-1}^{1} (-\varepsilon u'' + xu')\bar{u}e^{-x^2/2\varepsilon}dx + \|qu\|_\varepsilon^2 = \varepsilon\|u'\|_\varepsilon^2 + q\|u\|_\varepsilon^2$$

and hence

(3.8) $\qquad \varepsilon\|u'\|_\varepsilon^2 \le |\lambda + q|^2\|u\|_\varepsilon^2 + \|\Pi_\varepsilon u - \lambda u\|_\varepsilon\|u\|_\varepsilon \quad , \qquad \forall\lambda \in \mathbb{C}.$

Assuming $x \le 0$ we have

$$|u(x)|e^{-x^2/2\varepsilon} = -\int_0^1 \frac{d}{dt}\{(1-t)|u(x+t)|^2 e^{-(x+t)^2/2\varepsilon}\}dt$$

and hence we find in conjunction with (3.8):

(3.9) $\qquad |u(x)|^2 \le Ce^{x^2/2\varepsilon}\{\frac{1}{\varepsilon}\|u\|_\varepsilon^2 + \|u\|_\varepsilon\|u'\|_\varepsilon\} \le$

$$\le Ce^{x^2/2\varepsilon}\{\varepsilon^{-1}\|u\|_\varepsilon^2 + \varepsilon^{-\frac{1}{2}}\|u\|_\varepsilon^{3/2}\|\Pi_\varepsilon u - \lambda u\|_\varepsilon^{\frac{1}{2}}\} ,$$

where C is a generic constant depending on $\lambda$, but not on x, u and $\varepsilon$. For $x \ge 0$ we find the same inequality in an analogous way. This inequality yields suitable pointwise error estimates in an $\mathcal{O}(\sqrt{\varepsilon})$-neighbourhood of zero only, if $\|\cdot\|_\varepsilon$-estimates are known. Restricting the boundary value problem (3.1) to the subintervals $[-1,-\varepsilon^{\frac{1}{2}}M]$ and $[\varepsilon^{\frac{1}{2}}M,1]$, where M is larger than the largest zero of the Hermite polynomial $H_k$, we can prove uniform estimates with the aid of the maximum principle, cf. [3] or [4], and with the barrier function $H_k(x/\sqrt{2\varepsilon})\ \log(x/\sqrt{\varepsilon})$.

REFERENCES

[1] P.P.N. de Groen, *The nature of resonance in a singular perturbation problem of turning point type*, to appear (preprint: Memorandum 1977-003, Dept. of Math., Eindhoven Univ. of Technology, Eindhoven, The Netherlands, may 1977).

[2] R.C. Ackerberg & R.E. O'Malley, *Boundary layer problems exhibiting resonance*, Studies in Appl. Math. 49 (1970), p. 129-139.

[3] P.P.N. de Groen, *Spectral properties of second order singularly perturbed boundary value problems with turning points*, J. Math. Anal. Appl. 57 (1977), p. 119-149.

[4] R. Courant & D. Hilbert, *Methods of mathematical physics*, vol. II, Interscience Inc., New York, 1953.

*Differential Equations and Applications*
*W. Eckhaus and E.M. de Jager (eds.)*


# ASYMPTOTIC METHODS FOR RELAXATION OSCILLATIONS

J. Grasman*, M.J.W. Jansen** and E.J.M. Veling*

## 1. INTRODUCTION

The subject we deal with is presented in three parts. First we discuss the characteristics of *autonomous relaxation oscillations*. Examples and a survey of mathematical methods for those examples will be given. As a prototype of a relaxation oscillation is commonly used the periodic solution of the Van der Pol equation for large values of its parameter. This equation will be discussed in more detail.

In the second part discontinuous approximations of periodic solutions of the nonautonomous Van der Pol equation,

$$(1.1) \qquad \frac{d^2x}{dt^2} + \nu(x^2-1)\frac{dx}{dt} + x = b\cos t, \qquad \nu \gg 1,$$

will be made. For b independent of $\nu$ this problem exhibits subharmonic oscillations of period $2\pi m$, where m is an integer of order $O(\nu)$. The construction of the approximation brings about certain conditions for b and $\nu$. In the b,$\nu$-plane overlapping regions are found where these conditions are satisfied. In the domain of overlap two periodic solutions with different periods are possible which is in agreement with analytical and numerical results. The case m odd was analyzed in [9], here we will give a modified method covering the case m odd as well as m even.

Finally, we will investigate *mutually coupled relaxation oscillations* of Van der Pol's type. Apart from the large parameter $\nu$ a second, small parameter related to the weakness of the coupling is introduced. Applying asymptotic methods in both parameters we can approximate periodic solutions of the coupled system. The results for this class of problems may help us to

---

* Dept. of Applied Mathematics, Mathematical Centre, Amsterdam

** Dept. of Mathematics, Free University, Amsterdam.

understand interesting phenomena occuring in systems of interacting biologic oscillators. We mention certain forms of frequency entrainment leading to wave phenomena in systems of spatially distributed oscillators.

## 2. AUTONOMOUS RELAXATION OSCILLATIONS

In 1926 Balthasar van der Pol wrote his paper "On relaxation oscillations" [17], in which the periodic solution of the differential equation

$$\frac{d^2y}{dt^2} + \nu(y^2-1)\frac{dy}{dt} + y = 0 \tag{2.1}$$

was investigated for different values of $\nu$. Van der Pol remarked that the oscillations of (2.1) with $0 < \nu << 1$ differ considerably from those with $\nu >> 1$. In the first case the solution is a sinusoidal oscillation with a period close to $2\pi$, while for $\nu$ large the solution changes in time alternately very slow and very fast with a period propertional to $\nu$, see figure 2.1. For all $\nu > 0$ the oscillation has an amplitude $a(\nu)$ close to 2.
Van der Pol worked with a triode-circuit, in which fluctuations of the potential are described by (2.1). The parameter $\nu$ represents a time constant of the electrical system, the so-called time of relaxation. Since for $\nu >> 1$ the period is proportional to this parameter, he proposed to call the corresponding periodic solution a *relaxation oscillation*.

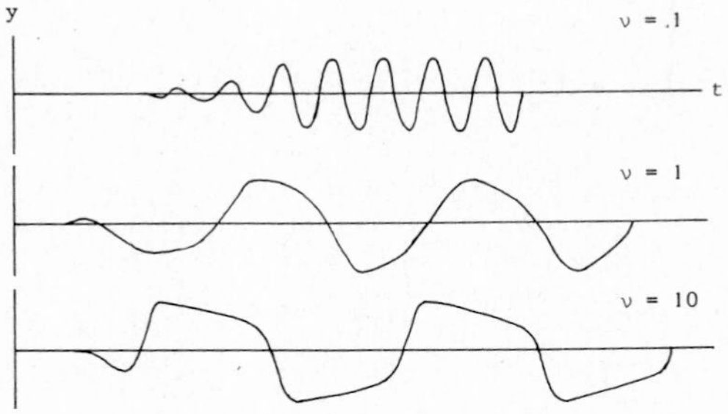


Fig. 2.1. Solutions of the Van der Pol equation for different values of the parameter.

In order to formulate a definition of relaxation oscillations for general autonomous systems of differential equations, we try to take some distance from Van der Pol's equation and introduce a new time-scale $\tau = t/\nu$ and a small parameter $\varepsilon = 1/\nu$. We consider the system of n equations

(2.2) $$p_i(\varepsilon)\frac{dx_i}{d\tau} = h_i(x_1,x_2,\ldots,x_n;\varepsilon), \qquad i = 1,2,\ldots,n,$$

where $p_i$ and $h_i$ are continuous functions in x and $\varepsilon$ for $0 < \varepsilon \le \varepsilon_0$ with $\varepsilon_0$ sufficiently small. It is assumed that

(2.3) $$\lim_{\varepsilon\to 0} h_i(x,\varepsilon) \quad \text{exists for bounded } x,$$

(2.4) $$p_1,p_2,\ldots,p_k \to 0, \quad p_{k+1},\ldots,p_n \to 1 \text{ as } \varepsilon \to 0 \quad (1\le k\le n).$$

Let the system (2.2) have a periodic solution with period $T(\varepsilon)$ and with closed trajectory $C(\varepsilon)$ in $\mathbb{R}^n$ such that

(2.5) $$x \in C(\varepsilon) \text{ implies } |x| \text{ is uniformly bounded for } 0 < \varepsilon \le \varepsilon_0$$

(2.6) $$\lim_{\varepsilon\to 0} T(\varepsilon) \text{ exists and is nonzero.}$$

<u>DEFINITION 2.1</u>. A periodic solution of (2.2) with period $T(\varepsilon)$ and closed trajectory $C(\varepsilon)$ satisfying (2.3) - (2.6) is called a relaxation oscillation if the converging sequences $\{x_{\varepsilon_q}\}$, $x_{\varepsilon_q} \in C(\varepsilon_q)$ with $\varepsilon_q \to 0$ as $q \to \infty$ form two nonempty sets $X_r$ and $X_s$:

(2.7) $$X_r = [\{x_{\varepsilon_q}\} \mid h_i(x_{\varepsilon_q},\varepsilon_q)/p_i(\varepsilon_q) \text{ converges as } q\to\infty \text{ for } i = 1,..,n],$$

(2.8) $$X_s = [\{x_{\varepsilon_q}\} \mid p_i(\varepsilon_q)/h_i(x_{\varepsilon_q},\varepsilon_q) \to 0 \text{ as } q \to \infty \text{ for some } i].$$

It is remarked that only nonlinear systems of the type (2.2) may exhibit relaxation oscillations. Furthermore, it is worth to mention that this definition does not provide a decisive answer on the stability of relaxation oscillations. At this point our definition does not concretize the existing vague idea that relaxation oscillations are asymptotically stable and, in case of forced oscillations, exhibit the phenomenon of frequency entrainment. There are Lyapunov stable (but not asymptotically stable) oscillations, which pass alternately the two characteristic phases of slow change and fast change in time as described by (2.7) and (2.8). It would lead to considerable confusion when these oscillations were termed differently. As an example of such oscillation we mention the periodic solutions of the

Volterra-Lotka equations for a certain range of the parameters. This system of equations has the form

$$\text{(2.9a)} \qquad \frac{dx_1^*}{dt^*} = x_1^*(-b+\beta x_2^*),$$

$$\text{(2.9b)} \qquad \frac{dx_2^*}{dt^*} = x_2^*(a-\alpha x_1^*).$$

Assuming that a << b we substitute

$$\text{(2.10ab)} \qquad x_1^* = \frac{b}{\alpha}x_1, \qquad t^* = \frac{1}{a}t,$$

$$\text{(2.10cd)} \qquad x_2^* = \frac{b}{\beta}x_2, \qquad a = \varepsilon b,$$

so that the system transforms into

$$\text{(2.11a)} \qquad \varepsilon\frac{dx_1}{dt} = x_1(-1+x_2),$$

$$\text{(2.11b)} \qquad \varepsilon\frac{dx_2}{dt} = x_2(\varepsilon-x_1), \qquad 0 < \varepsilon << 1.$$

A similar transformation can be made if a >> b. The system (2.11) has a one parameter family of periodic solutions with the equilibrium $(x_1,x_2) = (\varepsilon,1)$ as center point. In figure 2.2 we sketch the time-dependent behaviour of a periodic solution. In [8] it has been computed that the period satisfies

$$\text{(2.12)} \qquad T(\varepsilon) = (\mu-\theta) + \left(\frac{-1}{1-\theta} + \frac{1}{1-\mu}\right)\varepsilon \log\varepsilon \left[\frac{1}{1-\theta} - \frac{1}{1-\mu} + \frac{1}{1-\theta}\log\{(1-\theta)\log\frac{1}{\theta}\}\right.$$
$$\left. - \frac{1}{1-\mu}\log\{(\mu-1)\log\mu\} + I(\theta) + I(\mu)\right]\varepsilon + O(\varepsilon^2\log^2\varepsilon)$$

with

$$\theta - \log\theta = \mu - \log\mu = \sigma > 1, \qquad \theta < 1 < \mu,$$

$$I(\alpha) = \text{sign}(1-\alpha)\int_0^{-\ln\alpha}\left\{\frac{1}{(x+\alpha(1-e^x)} - \frac{1}{(1-\alpha)x}\right\}dx,$$

where $\theta = x_{2min}$, $\mu = x_{2max}$, and $\sigma = x_{1max} + O(\varepsilon\log\varepsilon)$.

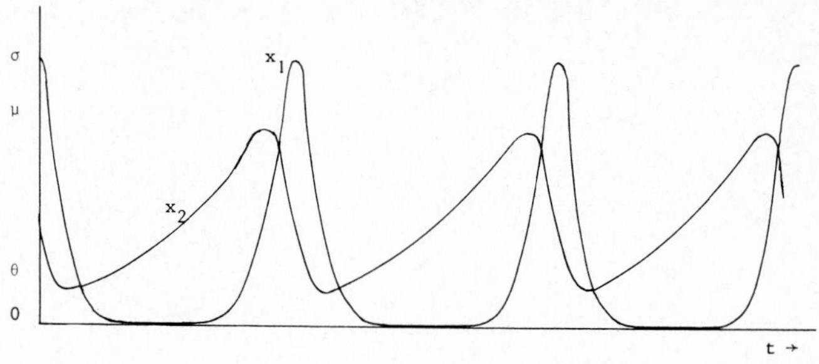


Fig. 2.2. A Volterra-Lotka system

For the proof of existence of periodic solutions of (2.2), mainly two types of methods are applied in literature. The first one is based on the *theorem of Poincaré-Bendixson*. For the Van der Pol equation and its generalizations this has been done a.o. by LEVINSON and SMITH [13], LASALLE [12] and PONZO and WAX [17]. This method only applies to second order autonomous systems. The other type of method uses *fixed point theorems*, see the work of CARTWRIGHT and LITTLEWOOD [4] for Van der Pol's equation with a forcing term and the theory of MISHCHENKO and PONTRYAGIN [16] for systems of the type (2.2).

An *asymptotic expansion* for the periodic solution of the Van der Pol equation with $\nu >> 1$ has been given by DORODNICYN [6]. Because of the changing behaviour of relaxations oscillations one has to apply methods different from those for almost linear oscillations as developed by BOGOLIUBOV and MITROPOLSKY [3]. Asymptotic methods for relaxation oscillations bear some resemblance with boundary layer techniques in fluid mechanics. One constructs local asymptotic approximations for each interval where the periodic solution has its own characteristic behaviour. The integration constants in these approximations are found by matching adjacent local approximations. In [2] the asymptotic method of Dorodnicyn has been modified at the point of matching. It was necessary to add a fifth local approximation in order to obtain a complete picture of the periodic solution of (2.1) for $\nu$ large. The following expressions for the amplitude and period were obtained.

(2.13) $$a(\nu) = 2 + \frac{1}{3}\alpha\nu^{-4/3} + \left(\frac{1}{3}b_1 - \frac{16}{27}\log\nu - \frac{1}{9} + \frac{2}{9}\log 2 - \frac{8}{9}\log 3\right)\nu^{-2} +$$

$$\left(\frac{1}{3}b_2 - \frac{2}{27}\alpha^2\right)\nu^{-8/3} + \left(\frac{1}{3}b_3 + \frac{104}{243}\alpha\log\nu - \frac{4}{27}\alpha b_1 - \frac{91}{486}\alpha + \frac{52}{81}\alpha\log 3 +\right.$$

$$- \frac{13}{81} \alpha \log 2)\nu^{-10/3} + o(\nu^{-10/3}),$$

(2.14) $$T(\nu) = (3-2 \log 2)\nu + 3\alpha\nu^{-1/3} - \frac{2}{3} \nu^{-1} \log \nu + \{\log 2 - \log 3 + 3b_1 +$$

$$- 1 - \log \pi - 2 \log Ai'(-\alpha)\}\nu^{-1} + o(\nu^{-1}),$$

with

$$\alpha = 2.33811, \qquad Ai'(-\alpha) = 0.70121,$$

$$b_1 = 0.17235, \qquad b_2 = 0.61778, \qquad b_3 = -0.55045.$$

## 3. THE VAN DER POL EQUATION WITH FORCING TERM

We study the Van der Pol equation with a periodic forcing term for large values of the parameter $\nu$:

(3.1) $$\frac{d^2x}{dt^2} + \nu(x^2-1) \frac{dx}{dt} + x = b \cos t.$$

For $b = 0$ the periodic solution is an autonomous relaxation oscillation as described in the preceding section. For $b > 0$ the system may have a periodic solution with a period m times the period of the driving term; this phenomenon is called *subharmonic entrainment*. The conditions on the values of $\nu$ and b under which this synchronization phenomenon occurs are derived in this section as the result of a formal approximation of the periodic solution by singular perturbation techniques with $1/\nu$ acting as a small parameter. These conditions bound regions in the $b,\nu$-plane where a solution with period $2\pi m$, might exist. It turns out that the regions, belonging to different values of m, have overlap; this is in agreement with results based on analytical-topological methods by LITTLEWOOD [14] and with numerical results by FLAHERTY and HOPPENSTEADT [7]. Making some modifications in the method of GRASMAN, VELING and WILLEMS [9] we will construct here a slightly different (lower order) approximation so that the case m even also can be included.

The synchronized solution of (3.1) with $b > 0$ can be considered as the sum the autonomous relaxation oscillation and a small harmonic oscillation. We shall make local approximations in different regions, see figure 3.1. To state the formal conditions for synchronization it is not necessary to consider more regions (as done in [9]). We just will use the knowledge that the jumps from ±1 to ∓2 take place in a time o(1). The method we use is related to Cole's treatment of the autonomous equation, see [5]. In the regions A and $\bar{A}$ we use two time scales, while in the regions B and $\bar{B}$ a stretching procedure with respect to the dependent variable is applied.

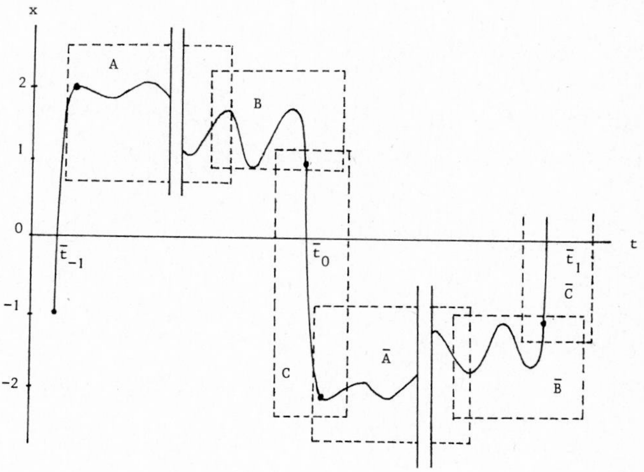


Fig. 3.1. Forced oscillations of the Van der Pol equation.

Region A

In this region the solution decays from the value 2 to 1 and has small amplitude oscillations of period $2\pi$. We apply the two-variable expansion procedure by introducing a second independent variable

(3.2) $$\tau = (t-t_0)/\nu.$$

We suppose that the solution can be written as

(3.3) $$x = x_0(\tau) + x_1(t,\tau)\nu^{-1} + o(\nu^{-1}).$$

Substituting (3.3) into equation (3.1) and letting $\nu \to \infty$ we find the limit equation

(3.4) $$(x_0^2-1)\left(\frac{\partial x_1}{\partial t} + \frac{\partial x_0}{\partial \tau}\right) + x_0 = b \cos t.$$

Solving (3.4) with respect to $x_1$ we find

$$(3.5)\qquad x_1(t,\tau) = \frac{b \sin t}{x_0^2-1} - \frac{p_1(\tau)t}{x_0^2-1},$$

with

$$p_1(\tau) \equiv (x_0^2-1)\frac{\partial x_0}{\partial \tau} + x_0.$$

As seen from (3.5) the term with $p_1$ is secular in the t variable, so we have to set $p_1(\tau) = 0$. Integration with respect to $\tau$ gives

$$(3.6)\qquad \log x_0 - \tfrac{1}{2}(x_0^2-1) = \tau.$$

In [9] it is demonstrated that the integration constants of (3.5) and (3.6) can be taken zero; the constant $t_0$ of (3.2) already accounts for these contributions. Thus,

$$(3.7)\qquad x_1(t,\tau) = \frac{b \sin t}{x_0^2(\tau)-1}.$$

When t approaches $t_0$, the behaviour of $x_0$ and $x_1$ is

$$(3.8ab)\qquad x_0 \approx 1+ \sqrt{t_0-t}\,\nu^{-1/2}, \qquad x_1 \approx \frac{b \sin t}{2\sqrt{t_0-t}}\,\nu^{1/2},$$

so for $t \to t_0$ the constructed solution behaves singular and looses its validity.

Region B

Let us suppose that for values of $t = t_0 + O(1)$ the solution is of the type

$$(3.9)\qquad x = 1 + U_0(t)\nu^{-1/2} + o(\nu^{-1/2}).$$

Substituting (3.9) into (3.1) and letting $\nu \to \infty$ we obtain

$$(3.10)\qquad 2U_0\frac{dU_0}{dt} + 1 = b \cos t,$$

so

$$(3.11)\qquad U_0(t) = \sqrt{b \sin t + t_0 - t + E_0}.$$

For $t_0 - t >> 1$, (3.12) behaves as

(3.13) $$U_0(t) \approx \sqrt{t_0 - t} + \frac{b \sin t + E_0}{2\sqrt{t_0 - t}} .$$

By inspection (3.13) matches (3.8) if $E_0 = 0$. Next we determine the point where the solution intersects the line $x = 1$. With an accuracy of $o(\nu^{-1/2})$ this will be at $t = \bar{t}_0$ satisfying $U_0(\bar{t}_0) = 0$ or

(3.14) $$t_0 - \bar{t}_0 = - b \sin \bar{t}_0 .$$

As we know from [9] at $t = \bar{t}_0$ the asymptotic solution jumps from 1 to the value -2 provided that

(3.15) $$\cos \bar{t}_0 < 1/b.$$

Region $\bar{A}$

Similar to the asymptotic solution of region A, we expand the solution in region $\bar{A}$ as

(3.16) $$x = y_0(\tau) + y_1(t,\tau)\nu^{-1} + o(\nu^{-1}), \qquad \tau = (t-t_0)/\nu$$

with

(3.17) $$\log(-y_0) - \tfrac{1}{2}(y_0^2-1) = \tau + (t_0-t_1)/\nu,$$

(3.18) $$y_1(t,\tau) = \frac{b \sin t}{y_0^2(\tau)-1} .$$

Since at the beginning of region $\bar{A}$ $(x,t) = (-2,\bar{t}_0)$ we derive from (3.16)-(3.18)

(3.19) $$t_1 = \bar{t}_0 - (\tfrac{3}{2} - \log 2)\nu + \tfrac{b}{2} \sin \bar{t}_0 + o(\nu^{-1}).$$

Region $\bar{B}$

For region $\bar{B}$ we have

(3.20) $$x = -1 - Z_0(t)\nu^{-1/2} + o(\nu^{-1/2}),$$

$$Z_0(t) = \sqrt{-b \sin t + t_1 - t} .$$

At region $\bar{B}$ the intersection with strip $x = 1 + o(\nu^{-1/2})$ takes place at $t = \bar{t}_1$ satisfying

$$(3.21) \qquad t_1 - \bar{t}_1 = b \sin \bar{t}.$$

At this point the solution jumps to $x = 2$ under the condition

$$(3.22) \qquad \cos \bar{t}_1 > -1/b.$$

At $x = 2$ the solution has been before at time $t = \bar{t}_{-1}$ according to (3.3) this was for

$$(3.23) \qquad \bar{t}_{-1} = t_0 - (\tfrac{3}{2} - \log 2)\nu + \tfrac{b}{2} \sin \bar{t}_{-1} + o(\nu^{-1}).$$

<u>Periodicity conditions</u>

We consider periodic solutions with period T being a multiple of $2\pi$ which intersect the line twice in a period. Such solution satisfies

$$(3.24) \qquad \bar{t}_1 - \bar{t}_{-1} = 2\pi m.$$

Let $2\delta(\nu)$ be the difference between the period $T_0(\nu)$ of the autonomous equation and the period T of the special solution, then

$$(3.25) \qquad 2\delta = T_0 - T = (3-2\log 2)\nu - 2\pi m + O(\nu^{-1/3}).$$

The system of equations (3.14), (3.19), (3.21), (3.23) and (3.24) can be reduced to

$$(3.26a) \qquad 3b(\sin \bar{t}_0 - \sin \bar{t}_1) = -4\delta,$$

$$(3.26b) \qquad b(\sin \bar{t}_0 + \sin \bar{t}_1) = -4(\bar{t}_1 - \bar{t}_0) + 4\pi m.$$

It turns out that the following change is suitable for the calculations

$$(3.27) \qquad \begin{aligned} \bar{t}_{-1} &= 2k_{-1}\pi + v_{-1} \\ \bar{t}_0 &= (2k_0+1)\pi + v_0 \\ \bar{t}_1 &= 2k_1\pi + v_1 \end{aligned}$$

with $-\pi < v_i \leq \pi$, $i = -1,0,1$. In view of the periodicity we have $v_{-1} = v_1$. For $b \leq 1$ equations (3.14) and (3.21) have a unique solution; for $b > 1$ we have to select the smallest root. In terms of $v_i$ the following condition has to be satisfied

$$(3.28) \qquad v_i + b \sin v_i > \sqrt{b^2-1} - \arccos(\tfrac{1}{b}) - \pi, \qquad i = 0,1.$$

Conditions (3.15) and (3.22) transform into

$$(3.29) \qquad \cos v_i > -1/b, \qquad i = 0,1.$$

The case m odd

For

$$(3.30a) \qquad 2k_1 - (2k_0+1) = (2k_0+1) - 2k_{-1} = m$$

$$(3.30b) \qquad v_0 = v_1$$

Equation (3.26b) is satisfied. Substitution in equation (3.26a) gives

$$(3.31) \qquad v_0 = v_1 = \arcsin\left(\frac{2\delta}{3b}\right),$$

so another natural restriction of the parameters is

$$(3.32) \qquad \left|\frac{2\delta}{3b}\right| \leq 1.$$

Conditions (3.29) are satisfied by (3.31), while (3.28) reads

$$(3.33) \qquad \arcsin \frac{2\delta}{3b} + \frac{2}{3} b > \sqrt{b^2-1} - \arccos(\tfrac{1}{b}) - \pi.$$

In the b,$\nu$-plane (3.32) and (3.33) determine the region, where a subharmonic solution with period $2\pi m$ with m odd may be expected, see figure 3.2. These are symmetric solutions satisfying $x(t) = -x(t-\frac{1}{2}T)$.

The case m even

If we set

$$(3.34a) \qquad 2k_1 - (2k_0+1) = m - 1,$$

(3.34b) $$(2k_0+1) - 2k_{-1} = m + 1,$$

the system (3.26) does not admit a solution of the type (3.30b). Besides the necessary condition (3.32) we find also by taking $v_{1\pi} = v_1 + \pi$ and applying the mean value theorem

$$\frac{\sin^2 v_0 - \sin^2 v_{1\pi}}{v_0 - v_{1\pi}} = \frac{16\delta}{3b^2}$$

the (solvability) condition

(3.35) $$\left|\frac{16\delta}{3b^2}\right| \leq 1.$$

In figure 3.2 we also give the region where a numerical solution for (3.26) was found that satisfied (3.28), (3.29), (3.32) and (3.35).

Some remarks

The regions in the b,$\nu$-plane corresponding with subharmonics of period $2\pi(2n-1)$ and $2\pi(2n+1)$ overlap. For a value of b and $\nu$ in the domain of overlap two different periodic solutions are possible depending on the initial values. The region corresponding with a subharmonic of period $4\pi n$ overlaps the two regions mentioned above in such a way that in a very narrow strip three subharmonics might exist. It is also possible to construct subharmonic solutions that intersect the line $x = 0$ 2q times ($q = 2,3,\ldots$) in one period, $T = (3-2\log 2)q\nu + O(1)$. This would lead to a system of 2q equations of the type (2.6). Such system can easily be reduced to a system of q equations in case of symmetric solutions with $x(t) = -x(t-\frac{1}{2}T)$. Finally, we remark that it is also possible to give sufficient conditions for solving the system (2.6) with m even. These conditions read

(3.36) $$1-\theta > \frac{4}{6}\left(\arccos\sqrt{\theta} + \frac{\pi}{2}\right) \quad \text{or} \quad 1-\theta > \frac{4}{b}\arcsin\sqrt{\theta}, \quad \theta = 2\delta/(3b).$$

In Littlewood's study [14] the amplitude b of the forcing term is of order $O(\nu)$. This leads to a same structure of subharmonic solutions with period $T = 2\pi(2n\pm 1)$ as found for b sufficiently large but independent of $\nu$. Littlewood states that for $b = \beta\nu$ with $\beta > 2/3$ only stable solutions of period $2\pi$ are found. Moreover, he signalized a what he called dipping phenomenon: the solution dips one or more times below the line $x = 1$ before jumping to the value $x = -2$ (a similar phenomenon may occur at $x = -1$).

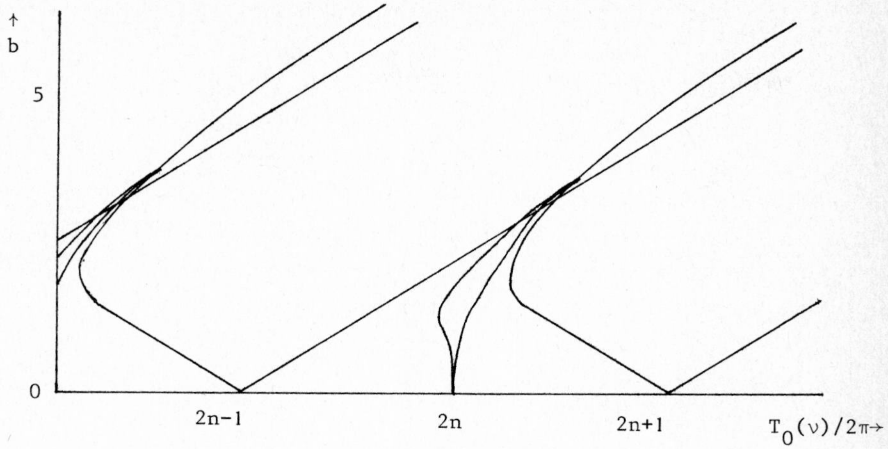


Fig. 3.2 Regions in the b,ν-plane with subharmonic solutions

## 4. WEAKLY COUPLED RELAXATION OSCILLATIONS

In this section we will present results on *coupled relaxation oscillations*. For the proofs of lemma's and theorems and for generalizations and further applications we refer to [10] and [11]. We consider a system of n coupled Van der Pol oscillators

$$\varepsilon \frac{d^2u_i}{dt^2} + (u_i^2-1)\frac{du_i}{dt} + u_i = \delta \sum_{j=1}^{n} h_{ij}(u_j), \qquad i = 1,2,\dots,n, \tag{4.1}$$

where $h_{ij}$ is a Lipschitz continuous function ($h_{ii}=0$) and $0 < \varepsilon,\ \delta \ll 1$. This system can be transformed into

$$\varepsilon \frac{du_i}{dt} = v_i - \frac{1}{3}u_i^3 + u_i \tag{4.2a}$$

$$\frac{dv_i}{dt} = -u_i + \delta \sum_{j=1}^{n} h_{ij}(u_j), \qquad i = 1,2,\dots,n. \tag{4.2b}$$

We also consider the degenerated system ($\varepsilon=0$)

$$0 = v_{0i} - \frac{1}{3}u_{0i}^3 + u_{0i} \tag{4.3a}$$

$$\frac{dv_{0i}}{dt} = -u_{0i} + \delta \sum_{j=1}^{n} h_{ij}(u_{0j}), \qquad i = 1,2,\dots,n. \tag{4.3b}$$

We introduce *formal discontinuous limit solutions* $(u_{0j}(t), v_{0j}(t))$ that satisfy (4.3) on regular arcs in the phase space with $|u_{0j}| > 1$. These arcs are connected by lines with $v_{0j}$ and $u_{0j}$ $(j \neq i)$ constant and with $u_{0i}$ varying from $\pm 1$ to $\mp 2$, denoting instantaneous jumps in $u_{0i}$. In the sequel it is assumed that at the end of a regular arc only one of the variables $u_j$ equals $\pm 1$. If such a sequence of connected arcs and lines forms a closed trajectory $Z_0^{(n)}$, then we have constructed a *formal discontinuous periodic limit solution*; its period $T_0^{(n)}$ is found by integration of (4.3b) over the regular arcs. For $n = 1$ we have the autonomous Van der Pol equation with $Z_0^{(1)}$ as sketched in figure 4.1. We denote the discontinuous periodic limit solution by

$$(4.4) \qquad u_{01}(t) = x_0(t), \quad v_{01}(t) = y_0(t).$$

Its period satisfies $T_0^{(1)} = 3 - 2 \log 2$.

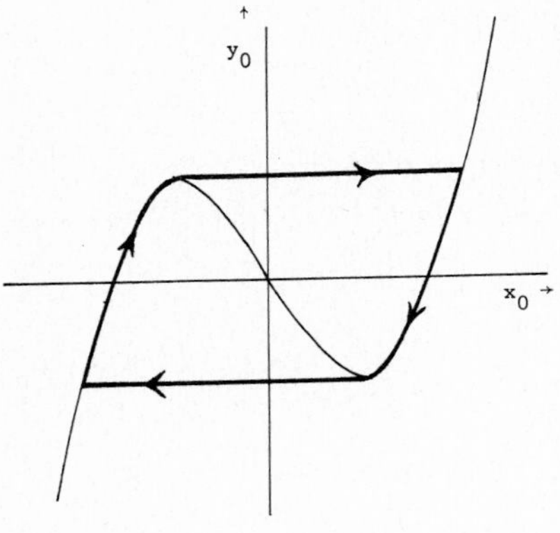


Fig. 4.1. The closed trajectory $Z_0^{(1)}$

Let us consider an $(n-1)$-dimensional surface P in $\mathbb{R}^{2n}$ satisfying (4.3a), $i = 1,\ldots,n$, being transversal to $Z_0^{(n)}$ at a point q of a regular arc. A formal discontinuous limit solution starting at P near q will return in a neighbourhood of q. In this way a mapping Q: P → P is defined. MISHCHENKO [15] proved the following theorem.

<u>THEOREM 4.1</u>. *Let the system* (4.2) *with degeneration* (4.3) *have a discontinuous periodic limit solution with closed trajectory* $Z_0^{(n)}$ *and let at a point* q *on a regular arc have a mapping* Q: P → P *be defined as above. If* Q *and its linearization at* q *are contracting, then the system* (4.2) *has for* $\varepsilon$ *suffi-*

*ciently small a periodic solution with period* $T_\varepsilon^{(n)}$ *and closed trajectory* $Z_\varepsilon^{(n)}$: $T_\varepsilon^{(n)} = T_0^{(n)} + O(\varepsilon^{2/3})$ *and* $Z_\varepsilon^{(n)} \to Z_0^{(n)}$ *as* $\varepsilon \to 0$.

From now on we focus our attention to the construction of $Z_0^{(n)}$. For $\delta$ sufficiently small we may write for the i-th component as a function of time

$$\text{(4.5a)} \qquad u_{0i}(t) = x_0(\phi_i(t)),$$

$$\text{(4.5b)} \qquad v_{0i}(t) = y_0(\phi_i(t)),$$

where $(x_0,y_0)$ is the discontinuous approximation of the autonomous Van der Pol equation. That is the i-th component of (4.3) runs the closed trajectory $Z_0^{(1)}$ of the autonomous Van der Pol equation in the limit $\varepsilon \to 0$. Substitution in (4.3b) gives

$$\text{(4.6)} \qquad \frac{d\phi_i}{dt} = 1 - \delta \sum_{j=1}^{n} h_{ij}[x_0(\phi_j)]/x_0(\phi_i), \qquad i = 1,2,\ldots,n.$$

The value of $\phi_i$ may be taken modulo $T_0^{(1)}$. Thus the problem is reduced to a system of n differential equations with function values on a n-dimensional torus. Let us set the discontinuities of $x_0(\phi)$ at $\phi = 0$ and $\phi = T_0^{(1)}/2$. We call a point $\alpha \in \mathbb{R}^n$ *regular* if the functions $u_{0i} = x_0(\alpha_i+t)$ are continuous in $t = 0$ and if they are discontinuous one at a time for $t > 0$.

LEMMA 4.1. *Let* $\phi(0) = \alpha$ *be regular. Then equation* (4.6) *has a unique solution* $\phi(t)$. *Moreover, for* $t$ *bounded (independent of* $\varepsilon$)

$$\text{(4.7)} \qquad \phi_i(t) = \alpha_i + t - \delta \sum_{j=1}^{n} \int_0^t \frac{h_{ij}[x_0(\alpha_j+\tau)]}{x_0(\alpha_i+\tau)}\, d\tau + O(\delta^2).$$

Let V be a (n-1)-dimensional plane orthogonal to $e = (1,1,\ldots,1)$ in $\mathbb{R}^n$. Let $\phi(0) \in V$ and let $T^*(\phi(0))$ be the time at which $\phi(t)$ returns in V. We consider the mapping $Q_V$ from V into V defined by

$$\text{(4.8)} \qquad \phi(0) \mapsto \phi(T^*(\phi(0))), \qquad T^*(\phi(0)) = T_0^{(1)} + \tau^*(\phi(0)),$$

see figure 4.2 for n = 2. Clearly, we must have $\tau^*(\phi(0)) = O(\delta)$. The first order approximation of $Q_V$ with respect to $\delta$ reads

$$\text{(4.9)} \qquad Q_V^{(0)}(\alpha) = \alpha + \delta G(\alpha) + \tau^*(\alpha)e,$$

where

$$G_i(\alpha) = -\sum_{j=1}^{n} \int_0^{T_0^{(1)}} \frac{h_{ij}[x_0(\tau)]}{x_0(\alpha_i-\alpha_j+\tau)}\, d\tau,$$

$$\tau^*(\alpha) = -\frac{\delta}{n} \sum_{i=1}^{n} G_i(\alpha).$$

Let this mapping have a fixed point $\tilde{\alpha}$: $Q_V^{(0)}(\tilde{\alpha}) = \tilde{\alpha}$.

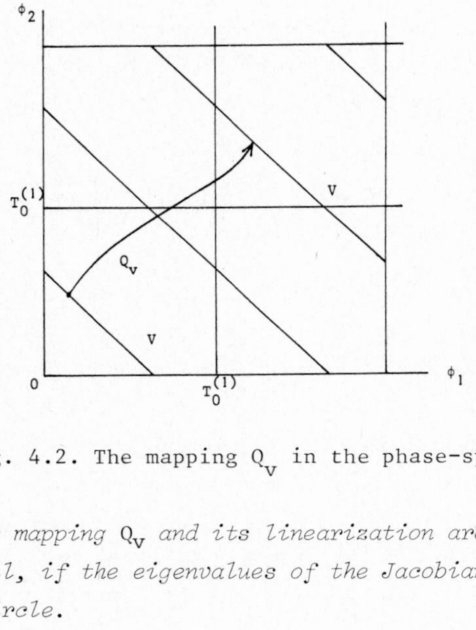


Fig. 4.2. The mapping $Q_V$ in the phase-space

<u>LEMMA 4.2</u>. *The mapping* $Q_V$ *and its linearization are contracting for* $\delta$ *sufficiently small, if the eigenvalues of the Jacobian of* $Q_V^{(0)}$ *are lying within the unit circle.*

For the problem (4.2) Mishchenko's theorem can be reformulated as follows.

<u>THEOREM 4.2</u>. *Let the system* (4.2) *have a formal discontinuous periodic limit solution satisfying* (4.3) *on the regular arcs and with jumps as prescribed. Let a point* q *on a regular arc be fixed point of a mapping* $Q_V$. *If the eigenvalues of the Jacobian of* $Q_V^{(0)}$ *are within the unit circle, then* (4.2) *has a periodic solution with period* $T_{\varepsilon,\delta}^{(n)}$ *and closed trajectory* $Z_{\varepsilon,\delta}^{(n)}$ *satisfying*

$$T_{\varepsilon,\delta}^{(n)} = T_0^{(1)} + \tau^*(\tilde{\alpha}) + O(\delta^2) + O(\varepsilon^{2/2})$$

*and*

$$Z^{(n)}_{\varepsilon,\delta} \to Z^{(n)}_{0} \quad \text{as} \quad \varepsilon,\delta \to 0.$$

It is noted that $\tau^*(\tilde{\alpha})$ is usually different from zero, which means that the period of the total system differs $O(\delta)$ from the period of the individual oscillators in decoupled state.

EXAMPLE 4.1. A system of Van der Pol oscillators on a circle with each oscillator only coupled with its direct neighbours may have the form

$$(4.10a) \qquad \varepsilon \frac{d^2u_1}{dt^2} + (u_1^2-1)\frac{du_1}{dt} + u_1 = \delta(u_n+u_1),$$

$$(4.10b) \qquad \varepsilon \frac{d^2u_i}{dt^2} + (u_i^2-1)\frac{du_i}{dt} + u_i = \delta(u_{i-1}+u_{i+1}), \qquad i = 2,\dots,n-1$$

$$(4.10c) \qquad \varepsilon \frac{d^2u_n}{dt^2} + (u_n^2-1)\frac{du_n}{dt} + u_n = \delta(u_{n-1}+u_1).$$

Considering the phase $\phi$ as a function of time and (discretized) position: $\phi = \phi(t,k\theta)$ with $\theta = 2\pi/n$, $k = 0,1,\dots,n-1$, we may find wave-type solutions satisfying

$$(4.11) \qquad n\{\phi(t,(k+1)\theta)-\phi(t,k\theta)\} = mT_0^{(1)}, \qquad k = 0,1,\dots,n-1,$$

with the circumference of the circle being m times the wave length. The conditions for theorem 4.2 are satisfied if $n \times m$ = odd (one oscillator jumps at a time) and if the eigenvalues of the Jacobian of $Q_v^{(0)}$ are within the unit circle, that is if

$$G'(\mu) - G'(-\mu) < 0$$

with

$$G(\mu) = -\int_0^{T_0^{(1)}} \frac{x_0(\tau)}{x_0(\tau+\mu)}\, d\tau, \qquad \mu = \frac{m}{n} T_0^{(1)}.$$

The results for this system of oscillators on a circle strongly resembles the behaviour of a model chemical reaction with diffusion taking place in a ring-shaped domain. AUCHMUTY and NICOLIS [1] analyzed this model reaction, first formulated by Prigogine, and found wave-type solutions similar to (4.11). Investigations on this model chemical reaction led to a better understanding of periodic phenomena in biochemistry and other areas of biology.

REFERENCES

[1] AUCHMUTY, J.F.G. & G. NICOLIS, *Bifurcation analysis of nonlinear reaction-diffusion equations - III, Chemical oscillations*, Bull. Math. Biol. 38 (1976), p. 325-350.

[2] BAVINCK, H. & J. GRASMAN, *The method of matched asymptotic expansions for the periodic solution of the Van der Pol equation*, Int. J. Nonlin. Mech. 9 (1974), p.421-434.

[3] BOGOLIUBOV, N.N. & I.A. MITROPOLSKY, *Asymptotic methods in the theory of nonlinear oscillations*, Gordon and Beach, New York (1961).

[4] CARTWRIGHT, M.L. & J.E. LITTLEWOOD, *On nonlinear differential equations of the second order*, Ann. Math. 54 (1951), p.1-37.

[5] COLE, J.D., *Perturbation methods in applied mathematics*, Blaisdell, Waltham, Mass. (1968).

[6] DORODNICYN, A.A., *Asymptotic solution of the Van der Pol equation*, Prikl. Mat. Mekh. 11 (1947), p.313-328. Am. Math. Soc. Transl., series 1, 4 (1962), p.1-23.

[7] FLAHERTY, J. & F.C. HOPPENSTEADT, *Frequency entrainment of a forced Van der Pol oscillator*, to appear in Studies in Appl.Math..

[8] GRASMAN, J. & E.J.M. VELING, *An asymptotic formula for the period of a Volterra-Lotka system*, Math. Biosci 18 (1973), p.185-189.

[9] GRASMAN, J., E.J.M. VELING & G.M. WILLEMS, *Relaxation oscillations governed by a Van der Pol equation with periodic forcing term*, SIAM J. Appl. Math. 31 (1976), p.667-676.

[10] GRASMAN, J. & M.J.W. JANSEN, *Mutually synchronized relaxation oscillators, as prototypes of oscillating systems in biology*, to appear.

[11] JANSEN, M.J.W., *Relaxation oscillators with weak additive coupling*, in preparation.

[12] LASALLE, J., *Relaxation oscillations*, Quart. J. Appl. Math. 7 (1949), p.1-19.

[13] LEVINSON, N. & O.K. SMITH, *A general equation for relaxation oscillations*, Duke Math. J. 9 (1942), p.382-403.

[14] LITTLEWOOD, J.E., *On nonlinear differential equations of the second order: the equation* $y'' - k(1-y^2)y' + y = b\mu k \cos(\mu t+\alpha)$ *for large* k, *and its generalizations*, Acta Math. 97 (1957), p.267-308.

[15] MISHCHENKO, E.F., *Asymptotic calculation of periodic solutions of systems of differential equations containing small parameters in the derivatives*, Izv. Akad. Nank. SSSR, Ser. Mat. 21 (1957), p.627-654. Am. Math. Soc. Transl., Ser. 2, 18 (1961), p.199-230.

[16] MISHCHENKO, E.F. & L.S. PONTYAGIN, *Differential equations with a small parameter attached to the higher derivatives and some problems in the theory of oscillation*, IRE Trans. on Circuit Theory CT-7 (1960), p.527-535.

[17] PONZO, P.J. & N. WAX, *On certain relaxation oscillations: Asymptotic solutions*, J. Soc. Industr. Appl. Math. 13 (1965), p.740-766.

[18] VAN DER POL, B., *On relaxation oscillations*, Phil. Mag. 2 (1926), p.978-992.

*Differential Equations and Applications*
*W. Eckhaus and E.M. de Jager (eds.)*
*©North-Holland Publishing Company (1978)*

# An Abstract Nonlinear Volterra Equation

J. A. Nohel[(1)]

University of Wisconsin, Madison, U.S.A.

## 1. Introduction.

In this lecture we discuss two problems concerning solutions of the abstract nonlinear Volterra equation

$$u(t) + b * Au(t) \ni F(t) \qquad (0 \le t \le T) , \tag{V}$$

where $T > 0$ is arbitrary, $A$ is an m-accretive (possibly multivalued) operator in a real Banach space $X$, the given kernel $b$ is a real absolutely continuous function on $[0,T]$, $b * g(t) = \int_0^t b(t-s)g(s)ds$ with the integral in (V) interpreted as the usual Bochner integral, and the given function $F \in W^{1,1}(0,T;X)$ where $W^{1,1}$ is the usual Sobolev space.

In problem I we treat the question of existence, uniqueness, dependence on data, and regularity of solutions of (V) on $[0,T]$ by means of a simple method developed recently and jointly with M. G. Crandall [8]; the results obtained for (V) generalize and simplify considerably earlier work on existence and uniqueness obtained by Barbu [2], [3], London [10], Gripenberg [9], all when $X = H$ is a real Hilbert space.

---

(1) Research sponsored by the United States Army under Grant No. DAAG29-77-G-0004 and by the National Science Foundation Grant No. MCS 75-21868.

Our method involves reducing the study of (V) to that of a related functional differential equation

$$(\mathrm{FDE}) \begin{cases} \dfrac{du}{dt} + Au \ni G(u) & (0 \le t \le T) \\ u(0) = x = F(0) , & \end{cases}$$

where $G : C([0,T]; \overline{D(A)}) \to L^1(0,T;X)$ is a given mapping, and developing a theory for (FDE) . Our results are also directly applicable to certain integrodifferential equations studied by McCamy [11] via a Galerkin argument which necessitates further restrictions.

We observe that if $b \equiv 1$ , equation (V) is equivalent to the evolution problem

$$(\mathrm{E}) \qquad \frac{du}{dt} + Au \ni F' , \quad u(0) = x = F(0) , \quad (0 \le t \le T) .$$

Our method of studying (FDE) consists of generalizing known results for (E) due primarily to Benilan [5] ; the latter are reviewed in Section 2 . We recall also that the initial-boundary value problem for a linear or nonlinear diffusion problem is a special case of (E) , and for this reason special cases of equation (V) (and also the related (FDE)) may be regarded as diffusion problems with a "memory".

In Problem II we discuss a result obtained jointly with P. Clement [6; Theorem 5] giving conditions under which the solution u of (V) preserves a closed convex cone in X ; this generalizes classical results on the positivity of solutions of the heat equation.

The results for Problems I and II are motivated and summarized in Section 3; for details of proofs the reader is referred to the appropriate literature. A model problem to which all of the theory of Problems I and II can be applied is given in [6; Section 4, Example 1].

2. Preliminaries on Evolution Equations. For further background and details of this section we refer the reader to [1], [5], [7].
Let $X$ be a real Banach space with norm $\|\cdot\|$. A mapping $A: X \to 2^X$ is called an operator in $X$; its domain $D(A) = \{x \in X : Ax \neq \phi\}$ and its range $R(A) = \bigcup\{Ax : x \in D(A)\}$; $A$ is single-valued if $Ax$ is a singleton. An operator $A$ in $X$ is accretive iff $J_\lambda = (I + \lambda A)^{-1}$ is a contraction in $X$ for $\lambda \geq 0$. It follows immediately: $A$ is accretive iff

$$\|(x_1 + \lambda y_1) - (x_2 + \lambda y_2)\| \geq \|x_1 - x_2\| \quad \text{for } y_i \in Ax_i \ (i = 1,2) . \tag{2.1}$$

An operator $A$ in $X$ is called m-accretive iff $A$ is accretive and $R(I + \lambda A) = X$ for $\lambda > 0$.

We shall be concerned with applying some known facts about the abstract evolution equation

$$\frac{dv}{dt} + Av \ni g , \quad v(0) = x \tag{Eg}$$

to the problem at hand. We assume throughout that $g \in L^1(0,T;X)$, $T > 0$.

Definition 2.1. A function $v : [0,T] \to X$ is a strong solution of (Eg) on $[0,T]$ if $v(0) = x$, $v \in C([0,T];X) \cap W^{1,1}(0,T;X)$, $v(t) \in D(A)$ a.e. on $[0,T]$ and there exists $w \in Av$ such that $w(t) \in Av(t)$ and $v'(t) + w(t) = g(t)$ a.e. on $[0,T]$.

Definition 2.2. $v : [0,T] \to X$ is a weak solution of (Eg) on $[0,T]$ if there is a sequence $\{(v_n, g_n)\}_{n=1}^{\infty} \subset C([0,T];X) \cap L^1(0,T;X)$ such that $v_n$ is a strong solution of $(Eg_n)$ on $[0,T]$ and $(v_n, g_n) \to (v,g)$ in $C([0,T];X) \cap L^1(0,T;X)$.

For our considerations we require a third concept of solution of (Eg), namely the notion of integral solution. First, let $[\ ,\ ]_\lambda : X \times X \to \mathbb{R}$ be defined for $\lambda \neq 0$ by

$$[x,y]_\lambda = \frac{1}{\lambda}(\|x + \lambda y\| - \|x\|) ,$$

which is a nondecreasing function of $\lambda$. Define

$$[x,y]_+ = \lim_{\lambda \downarrow 0} [x,y]_\lambda = \inf_{\lambda > 0} [x,y]_\lambda$$

$$[x,y]_- = \lim_{\lambda \uparrow 0} [x,y]_\lambda = \sup_{\lambda < 0} [x,y]_\lambda .$$

Thus $\|x + \lambda y\| \geq \|x\|$ for $\lambda \geq 0$ iff $[x,y]_+ \geq 0$, so that A is accretive iff

(2.2) $$[x_1 - x_2 , y_1 - y_2]_+ \geq 0 \text{ for } y_i \in Ax_i .$$

Definition 2.3. $v : [0,T] \to X$ *is an integral solution of* (Eg) *on* $[0,T]$ *if* $v \in C([0,T];X$ *and*

(2.3) $$\| v(t) - x\| - \| v(s) - x\| \leq \int_s^t [v(\alpha) - x , g(\alpha) - y]_+ \, d\alpha$$

*for* $t > s$, $(t,s) \in [0,T]$, $x \in D(A)$ *and* $y \in Ax$. We note since $|[x,y]_+| \leq \|y\|$, and since $g \in L^1(0,T;X)$, the integral in (2.3) is well defined. A straightforward calculation, see [7], shows that the notion of integral solution only makes sense when A is accretive. We shall apply the following result on existence, uniqueness, dependence on data, and regularity about integral solutions of (Eg) due to Benilan [5].

Theorem A. *If* A *is* m-*accretive,* $x \in \overline{D(A)}$, *and* $g \in L^1(0,T;X)$ *then* (Eg) *has a unique integral solution* $v \in C([0,T];\overline{D(A)})$ *on* $[0,T]$, *and if* $v, \hat{v}$ *are integral solutions of* (Eg), $(\hat{Eg})$ *on* $[0,T]$ *corresponding to initial* values $x, \hat{x}$ respectively then

(2.4) $$\| \hat{v}(t) - v(t)\| \leq \|x - \hat{x}\| + \int_0^t \|\hat{g}(\sigma) - g(\sigma)\| d\sigma , \quad 0 \leq t \leq T .$$

*Moreover, if* $g \in BV([0,T];X)$ *and* $x \in D(A)$, *then*

(2.5) $$\|v(\xi) - v(\eta)\| \leq |\xi - \eta| \{\|g(0^+) - y\| + \operatorname{var}(g : [0,t])\}$$

*for* $y \in Ax$ , *and* $0 \le \xi, \eta \le t$ , $t \in [0,T]$ . *In particular, the integral solution* $v$ *is Lipschitz continuous*. *If, in addition,* $X$ *is reflexive, then* $v$ *is a strong solution of* (Eg) *on* $[0,T]$ .

3. *Summary of Results*. (I) We shall reduce the study of existence and uniqueness of solutions of the nonlinear Volterra equation (V) on $[0,T]$ to studying the abstract functional differential equation

$$\text{(FDE)} \begin{cases} \dfrac{du}{dt} + Au \ni G(u) & (0 \le t \le T) \\ u(0) = x , \end{cases}$$

where $A$ is a given m-accretive operator on $X$ , and where $G$ is a given mapping

$$G : C([0,T]; \overline{D(A)}) \to L^1(0,T;X) .$$

Let $v = H(g)$ denote the unique integral solution of (Eg) .
*A solution of* (FDE) *is by definition a function* $u \in C([0,T]; \overline{D(A)})$ *such that* $u = H(G(u))$ . By analogy with Definition 2.1, we say that $u$ is a *strong solution of* (FDE) on $[0,T]$ if $u(0) = x$ , $u \in W^{1,1}(0,T;X) \cap C([0,T]; \overline{D(A)})$ and if $u'(t) + Au(t) \ni G(u)(t)$ a.e. on $[0,T]$ .

Let $b \in L^1(0,T;\mathbb{R})$ , $F \in L^1(0,T;X)$ . *We shall say that* $u$ *is a strong solution of the Volterra equation* (V) *on* $[0,T]$ *if* $u \in L^1(0,T;X)$

*and if there exists* $w \in L^1(0,T;X)$ *such that* $w(t) \in Au(t)$ *and* $u(t) + b * w(t) = F(t)$ a.e. *on* $[0,T]$. One can establish the following equivalence between strong solutions of (FDE) with a particular $G$ and strong solutions of (V) :

*Proposition 1.* *Let* $b \in AC([0,T];\mathbb{R})$, $b' \in BV([0,T];\mathbb{R})$, $F \in W^{1,1}(0,T;X)$ *and* $b(0) = 1$. *Let* $u$ *be a strong solution of* (V) *on* $[0,T]$. *Then* $u$ *is a strong solution of* (FDE) *on* $[0,T]$ *with the identifications* :

$$(3.1)\quad \begin{cases} \text{(i)} & G(u)(t) = f(t) - r * f(t) - a(0)u(t) - r(t)x + \int_0^t u(t-s)dr(s) \\ \text{(ii)} & f(t) = F'(t) \\ \text{(iii)} & x = F(0) \\ \text{(iv)} & a = b' \\ \text{(v)} & r \in L^1(0,T;\mathbb{R}) \text{ is defined by } r + a * r = a . \end{cases}$$

*Conversely, let* $r \in BV([0,T];\mathbb{R})$, $f \in L^1(0,T;X)$, $x \in \overline{D(A)}$ *and* $G$ *be given by* (3.1) (i). *Let* $u$ *be a strong solution of* (FDE) *on* $[0,T]$. *Then* $u$ *is a strong solution of* (V) *on* $[0,T]$, *where*

$$(3.2)\quad \begin{cases} \text{(i)} & F(t) = x + \int_0^t f(s)ds \\ \text{(ii)} & b(t) = 1 + \int_0^t a(s)ds \\ \text{(iii)} & a + a * r = r . \end{cases}$$

We remark that if $b(t) \equiv 1$ and $F \in W^{1,1}(0,T;X)$, then the Volterra equation (V) is equivalent to the evolution equation (Eg) where $g = F'$ and where the initial value $x = F(0)$.

The proof of Proposition 1 is straightforward. The assumptions on b and F permit differentiation a.e. on $[0,T]$ of a strong solution u of (V). The differentiated equation is then "solved" for Au by means of the resolvent kernel r associated with $a = b'$, see (3.1) (v). A known result [4] yields that $a \in BV([0,T];\mathbb{R})$ implies that $r \in BV([0,T];\mathbb{R})$, a fact which is used in arriving at the formula (3.1) (i) for G(u). The converse is proved by reversing the steps. A part of Proposition 1 which motivates our approach is contained in MacCamy [11] who, however, then studied (FDE) by an entirely different approach.

We remark that here we have chosen to define the resolvent kernel by (3.1) (v), rather than by $r + a * r = -a$ as was done in [8]. This is more convenient for Theorem 5 below, and only causes a change of signs in the formula (3.1) (i) of some of the terms in G(u)(t). Recall that if r is defined by (3.1) (v), then the solution of the linear Volterra equation $w + a * w = v$ is given by $w = v - r * v$, while for the alternate definition of r, w would be given by $w = v + r * v$, as was used in [8].

We next use Benilan's theorem about solutions of (Eg) to obtain some general results concerning existence, uniqueness, dependence on data, and regularity of solutions of (FDE) of independent interest and use them to deduce corresponding results about solutions of (V).

Theorem 1. *Assume that* $A$ *is* *m-accretive*, $x \in \overline{D(A)}$, *and let* $G : C([0,T];\overline{D(A)}) \to L^1(0,T;X)$ *satisfy*

$$(3.3)\quad \begin{cases} \|G(u)-G(v)\|_{L^1(0,t;X)} \le \int_0^t \gamma(s)\|u-v\|_{L^\infty(0,s;X)}\,ds \\ \text{for some } \gamma \in L^1(0,T;\mathbb{R}^+),\ 0 \le t \le T,\ \text{and } u,v \in C([0,T];\overline{D(A)}). \end{cases}$$

*Then* (FDE) *has a unique solution* $u \in C([0,T];\overline{D(A)})$ *on* $[0,T]$.

We remark that assumption (3.3) implies that the value of $G(u)$ at $t \in [0,T]$ depends only on the restriction of $u$ to $[0,t]$. The idea of the proof is very simple. Let $v = H(g)$ denote the unique integral solution of (Eg) on $[0,T]$, $g \in L^1(0,T;X)$. We seek a fixed point of the map $K: C([0,T];\overline{D(A)}) \to C([0,T];\overline{D(A)})$ defined by $K(u) = H(G(u))$. By property (2.4) of integral solutions

$$\|K(u)(t)-K(v)(t)\| \le \int_0^t \|G(u)(s)-G(v)(s)\|\,ds \qquad (0 \le t \le T),$$

for $u,v \in C([0,T];\overline{D(A)})$, $u(0)=v(0)=x$. Applying assumption (3.3) it is now an easy matter to show that $K^j$ is a strict contraction on $C([0,T];\overline{D(A)})$ for $j$ sufficiently large, so that the map $K$ has a unique fixed point. For details see [8].

Under further assumptions one can apply the second part of Benilan's theorem to obtain greater regularity of solutions of (FDE).

*Theorem 2.* *In addition to the assumptions of Theorem 1 assume that there is a function* $k : [0,\infty) \to [0,\infty)$ *such that*

$$(3.4)\qquad \begin{cases} \text{var}\,(G(u) : [0,t]) \le k(R)(1 + \text{var}\,(u : [0,t])) \\ \text{and} \quad \|G(u)(0^+)\| \le k(R) \qquad (0 \le t \le T)\,, \end{cases}$$

*whenever* $u \in C([0,T]; \overline{D(A)})$ *is of bounded variation and* $\|u\|_{L^\infty(0,T;X)} \le R$ . *If* $x \in D(A)$ , *then the solution* $u$ *of* (FDE) *is Lipschitz continuous on* $[0,T]$ . *If* $X$ *is also reflexive, then the solution* $u$ *is a strong solution of* (FDE) *on* $[0,T]$ .

For the proof of Theorem 2 one defines $u_0 : [0,T] \to X$ by $u_0(t) = x$ and $u_{n+1} = K(u_n) = H(G(u_n))$ , $n = 0,1,\cdots$ . These iterates converge uniformly and are uniformly bounded on $[0,T]$ . By Benilan's theorem and assumption (3.4) one shows that there exists a constant $c > 0$ such that

$$\text{var}\,(u_{n+1} : [0,t]) \le c(1 + \int_0^t \text{var}\,(u_n : [0,s])\,ds)$$

for $0 \le t \le T$ , so that $\text{var}\,(u_{n+1} : [0,t]) \le c \exp(ct)$ . Thus $\{\text{var}(u_n : [0,T])\}$ and by (3.4) $\{\text{var}(G(u_n)) : [0,T]\}$ are both bounded, and $\{u_n\}$ , and hence also $u = \text{unif lim}\, u_n$ , is Lipschitz continuous on $[0,T]$ . For more details see [8] .

Finally, the solution $u$ of (FDE) depends on the data $A$ , $G$ , $x$ in the following sense:

Theorem 3. *Let the assumptions of Theorem 1 be satisfied. Let m-accretive operators* $A_n$ *in* $X$, *mappings* $G_n : C([0,T];X) \to L^1(0,T;X)$, *and* $x_n \in \overline{D(A)}$ *be given for* $n = 1,2,\cdots$. *Assume that the inequality* (3.3) *holds for* $G$ *replaced by* $G_n$, *with the same* $\gamma$, *for* $n = 1,2,\cdots$, *and* $u, v \in C([0,T];\overline{D(A)})$. *For* $u \in C([0,T];\overline{D(A)})$ *assume that* $\lim_{n\to\infty} G_n(u) = G(u)$ *in* $L^1(0,T;X)$, $\lim_{n\to\infty} x_n = x \in \overline{D(A)}$, *and*

$$\lim_{n\to\infty} (I + \lambda A_n)^{-1} z = (I + \lambda A)^{-1} z \qquad (z \in X,\ \lambda > 0). \tag{3.5}$$

*Let* $u_n \in C([0,T];\overline{D(A_n)})$ *be solutions of* (FDE) *on* $[0,T]$ *with* $A$ *replaced by* $A_n$, $G$ *replaced by* $G_n$, $x$ *replaced by* $x_n$, *and let* $u \in C([0,T];\overline{D(A)})$ *be the solution of* (FDE) *on* $[0,T]$. *Then* $\lim_{n\to\infty} u_n = u$ *in* $C([0,T];X)$.

The proof of Theorem 3 follows from the observation that under our assumptions the mapping $K(A,x,G)(u) = H(A,x,G(u))$ of Theorem 1 has the property that in the iterate $K^j$, which is a strict contraction for some $j$, both $j$ and the contraction constant depend only on the function $\gamma$ of (3.3), and the latter is assumed to be uniform in $n$; for details see [8].

We shall next apply Theorems 1, 2, 3 to study the nonlinear Volterra equation (V). If $b$ and $F$ in (V) satisfy the assumptions of Proposition 1, it follows from the definition of $G$ in (3.1) (i) that

$$\| G(u)(t) - G(v)(t) \| \le (|r(0^+)| + \mathrm{var}\,(r : [0,t]))\, \| u - v \|_{L^\infty(0,t;X)},$$

where $r$ is the resolvent kernel corresponding to $b' = a$ (recall that $a \in BV([0,T];\mathbb{R}) \Longrightarrow r \in BV([0,T];\mathbb{R})$) . Thus assumption (3.3) of Theorem 1 is satisfied with

$$\gamma(s) = |r(0)| + \text{var}\,(r : [0,s]) .$$

Moreover, if $f = F' \in BV([0,T];X)$ , (3.1) (i), (ii), (iii) imply

$$\text{var}\,(G(u) : [0,t]) \le C(1 + \text{var}\,(u : [0,t])) \qquad (0 \le t \le T) ,$$

and $\|G(u)(0^+)\| \le C$ , where $C$ is a constant depending on $F(0)$ , $F'(0^+)$ , $\text{var}\,(F' : [0,T])$ , $r(0^+)$ , and $\text{var}\,(r : [0,T])$ ; thus assumption (3.4) of Theorem 2 is satisfied.

Let $\lambda > 0$ and define the Yosida approximation $A_\lambda$ of the m-accretive operator $A$ on $X$ by

$$A_\lambda = \frac{1}{\lambda}(I - J_\lambda) , \qquad J_\lambda = (I + \lambda A)^{-1} .$$

$A_\lambda : X \to X$ is Lipschitz continuous, so a simple contraction argument shows that the approximating problem

$$(V_\lambda) \qquad u_\lambda + b * A_\lambda u_\lambda = F$$

has a unique strong solution $u_\lambda$ on $[0,T]$ , under the assumptions :

$b \in L^1(0,T;\mathbb{R})$, and $F \in L^1(0,T;X)$. By Proposition 1 $u_\lambda$ is a strong solution of

$$(FDE_\lambda) \qquad \frac{du_\lambda}{dt} + A_\lambda u_\lambda = G(u_\lambda), \quad u_\lambda(0) = F(0).$$

One also has $\lim_{\lambda \downarrow 0} (I + \mu A_\lambda)^{-1} z = (I + \mu A)^{-1} z$, for $\mu > 0$, $z \in X$.

These considerations lead to the following result about solutions of (V).

Theorem 4. *Let $A$ be an m-accretive operator on $X$; let $b, F$ satisfy the assumptions of Proposition 1, and let $F(0) \in \overline{D(A)}$. Let $u_\lambda$ be the solution of $(V_\lambda)$ on $[0,T]$. Then $\lim_{\lambda \downarrow 0} u_\lambda = u$ in $C([0,T];X)$ where $u$ is the solution of (FDE) on $[0,T]$ of Theorem 1 with the identifications (3.1). If, moreover, $F' \in BV([0,T];X)$ and $F(0) \in D(A)$, then the solution $u$ of (FDE) is Lipschitz continuous on $[0,T]$. If $X$ is also reflexive $u$ is a strong solution of (V) on $[0,T]$.*

We remark that if the Volterra equation (V) has a strong solution $u$ on $[0,T]$ under the assumptions of Theorem 2, then from Theorem 2 and Proposition 1, $\lim_{\lambda \downarrow 0} u_\lambda = u$ in $C([0,T]:X)$ exists, where $u_\lambda$ are the strong solution of the approximating equation $(V_\lambda)$. However, under our assumptions the solutions $u_\lambda$ of $(V_\lambda)$ converge to a limit $u$ as $\lambda \downarrow 0$, whether or not (V) has a strong solution. For this reason we refer to the solution $u$ of (V) of Theorem 4 as the *generalized solution* of V on $[0,T]$.

A precise estimate giving the dependence of the generalized solution u of (V) on the data b and F is established in Theorem 5 of [8].

We remark that the assumption $b(0) = 1$ in Theorem 4 and Proposition 1 is no loss of generality, provided $b(0) > 0$. For if $b(0) > 0$, defining $\tilde{b} = (b(0))^{-1} b$ and $\tilde{A} = b(0) A$ one has $b * Au = \tilde{b} * \tilde{A} u$.

We point out also that our method can be used to study the nonconvolution Volterra equation

$$u(t) + \int_0^t b(t,s) A u(s) ds \ni F(t) \qquad (0 \le t \le T),$$

where A and F are as in Theorem 4, provided the kernel b, which is defined on the region $\{(t,s) : 0 \le s \le t \le T\}$, is sufficiently smooth and $b(t,t) > 0$. The technique for doing this is outlined in [8], and is carried out in detail by C. Rennolet [12].

(II) Concerning the linear or nonlinear diffusion equation with appropriate boundary and initial conditions, a special case of (V) with $b \equiv 1$ and $F \in W^{1,1}(0,T;X)$, there are classical results concerning the positivity of solutions. We present a general result of this nature for the solution of (V). Let A be an m-accretive operator on X and let $P \subseteq X$ be a closed convex cone satisfying the condition

$$J_\lambda P = (I + \lambda A)^{-1} P \subseteq P \qquad (\lambda > 0). \tag{3.6}$$

Consider the approximating equation $(V_\lambda)$ which, as pointed out in the remarks leading up to Theorem 4, has a unique strong solution $u_\lambda$ on $[0,T]$ for $\lambda > 0$ under the assumption $b \in L^1(0,T;\mathbb{R})$, $F \in L^1(0,T;X)$. Equation $(V_\lambda)$ can be written in the equivalent form

$$u_\lambda + \frac{1}{\lambda} b * u_\lambda = F + \frac{1}{\lambda} b * J_\lambda u_\lambda \qquad (\lambda > 0) . \tag{3.7}$$

It is easy to give conditions on $b$ and $F$ which insure that solution $u_\lambda$ of $(V_\lambda)$ satisfies $u_\lambda(t) \in P$ a.e. on $[0,T]$.

For $b \in L^1(0,T;\mathbb{R})$ define the function $\rho_\mu : [0,T] \to \mathbb{R}$ to be the unique solution of the scalar resolvent equation

$$\rho(t) + \mu b * \rho(t) = b(t) \qquad (0 \le t \le T; \ \mu \ge 0) ; \tag{$\rho_\mu$}$$

then $\rho_\mu = \rho(\cdot,\mu) \in L^1(0,T;\mathbb{R})$. We shall assume that

$$\text{For every } \mu \ge 0 \quad \rho(t,\mu) \ge 0 \text{ a.e. on } [0,T] . \tag{3.8}$$

We also define the function $s_\mu : [0,T] \to \mathbb{R}$ to be the unique solution of the scalar equation

$$s(t) + \mu b * s(t) = 1 \qquad (0 \le t \le T; \ \mu \ge 0) . \tag{$s_\mu$}$$

If $b \in L^1(0,T;\mathbb{R})$, $s_\mu = s(\cdot,\mu) \in L^1(0,T;\mathbb{R})$. It is readily verified [6] that

$$s(t,\mu) = 1 - \mu \int_0^t \rho(\xi,\mu)d\xi \qquad (0 \le t \le T),$$

so that $s(t,\mu)$ is absolutely continuous on $[0,T]$. We shall assume that

(3.9) for every $\mu \ge 0$ $s(t,\mu) \ge 0$ on $[0,T]$.

The following result partly implicit in the literature is established in [6].

Proposition 2. *Let* $b \in L^1(0,T;\mathbb{R})$.

(i) *If* $b \in C((0,T);\mathbb{R})$, $b(t) > 0$, *and* $\log b(t)$ *is convex on* $(0,T)$, *then* (3.8) *is satisfied*.

(ii) *If* $b(t)$ *is nonnegative and nonincreasing on* $(0,T)$ *then* (3.9) *is satisfied*.

Thus, for example, if $b \in L^1(0,T;\mathbb{R})$ and $(-1)^k b^{(k)}(t) \ge 0$ $(0 < t < T;\ k = 0,1,\cdots)$, then both (3.8) and (3.9), as well as the assumptions of Theorem 4 concerning $b$, if $0 < b(0) < \infty$ are satisfied. For other examples see [6].

We shall also assume that:

(3.10) $\left\{\begin{array}{l} \text{for every } \mu > 0 \text{ the unique solution } z_\mu \text{ of the} \\ \text{linear equation} \\ \quad z(t) + \mu b * z(t) = F(t) \qquad t \in [0,T] \quad \text{a.e.} \\ \text{satisfied } z_\mu(t) \in P \quad \text{a.e.} \quad \text{on} \quad [0,T] \;. \end{array}\right.$

It can be shown [6] that if $F \in W^{1,1}(0,T;X)$, $F(0) \in P$ and $F'(t) \in P$ a.e. on $[0,T]$, then (3.10) is satisfied, provided (3.9) holds. If $F(t) = u_0 + b * h(t)$, $h \in L^1(0,T;X)$, then (3.10) is satisfied provided $u_0 \in P$, $h(t) \in P$ a.e. on $[0,T]$, and (3.9) holds.

Returning to the approximating equation $(V_\lambda)$ written in the form (3.7), let $\rho(\cdot,\frac{1}{\lambda}) \in L^1(0,T;\mathbb{R})$ be the unique solution of the scalar equation $(\rho_\mu)$ with $\mu$ replaced by $\frac{1}{\lambda}$, and for $F \in L^1(0,T;X)$ define $F_\lambda : [0,T] \to X$ by

$$F_\lambda(t) = F(t) - \frac{1}{\lambda} \int_0^t \rho(t-\sigma, \frac{1}{\lambda})\, F(\sigma)d\sigma \qquad (0 \le t \le T) \;.$$

By standard theory of linear Volterra equations $F_\lambda(\cdot) \in L^1(0,T;X)$ is the unique solution of the linear equation in assumption (3.10) with $\mu$ replaced by $\frac{1}{\lambda}$. It follows also by this theory that the approximating equation $(V_\lambda)$ is equivalent to the equation

(3.11) $$u_\lambda = F_\lambda + W_\lambda(u_\lambda) \qquad (\lambda > 0)$$

where

$$W_\lambda(q)(t) = \frac{1}{\lambda} \int_0^t \rho(t - \sigma, \frac{1}{\lambda}) J_\lambda(q)(\sigma)d\sigma .$$

These considerations lead to the following result established in [6, Theorem 5] which extends to solutions of $(V_\lambda)$ and $(V)$ the classical theorems about positivity of solutions of the heat equation.

Theorem 5. *Let* $b \in L^1(0,T;\mathbb{R})$ , $F \in L^1(0,T;X)$ , $A$ *be an m-accretive operator on* $X$ , *and let assumptions* (3.6) , (3.8) , (3.10) *be satisfied. Then* $u_\lambda$ , *the strong solution of* $(V_\lambda)$ , *satisfies* $u_\lambda(t) \in P$ a.e. *on* $[0,T]$ . *Consequently, if also the conditions of Theorem* 4 *are satisfied, then* $u$ , *the generalized solution of equation* $(V)$ , *also satisfies* $u(t) \in P$ a.e. *on* $[0,T]$ .

For the proof of Theorem 5 let $w_\lambda = u_\lambda - F_\lambda$ and write (3.11) in the equivalent form

$$w_\lambda = W_\lambda(w_\lambda + F_\lambda) .$$

Noting that $W_\lambda$ maps $L^1(0,T;X)$ into itself and that by the contraction property of $J_\lambda$ (recall $A$ is m-accretive), we prove (see [6] ) that some iterate $W_\lambda^{n_\lambda}$ of $W_\lambda$ is a strict contraction for $n_\lambda$ sufficiently large. Then $u_\lambda = F_\lambda + \lim_{n\to\infty} W_\lambda^n(u_0)$ , $u_0 \in L^1(0,T;X)$ , is the unique solution of (3.11) , and if $u_0(t) \in P$ a.e. on $[0,T]$ , our assumptions guarantee that the same holds for $u_\lambda(t)$ .

## References


1. V. Barbu, Nonlinear Semigroups and Differential Equations in Banach Spaces. Noordhoff International Publishing, 1976.
2. V. Barbu, Nonlinear Volterra equations in Hilbert space. SIAM J. Math. Anal. 6 (1975), 728-741.
3. V. Barbu, On a nonlinear Volterra integral equation on a Hilbert space. SIAM J. Math. Anal. 8 (1977), 345-355.
4. R. Bellman and K. L. Cooke, Differential-Difference Equations. Academic Press, 1963.
5. Ph. Benilan, Equations d'évolution dans une espace de Banach quelconque et applications. Thèse de Doctorat d'État, Univ. de Paris (Orsay), 1972.
6. P. Clément and J. A. Nohel, Abstract linear and nonlinear Volterra equations preserving positivity. Math. Res. Center, Univ. of Wisconsin, Tech. Summary Report #1716, 1977.
7. M. G. Crandall, An introduction to evolution governed by accretive operators. Dynamical Systems, Vol. 1, Academic Press, 1976, pp. 131-165.
8. M. G. Crandall and J. A. Nohel, An abstract functional differential equation and a related nonlinear Volterra equation. Israel Math. J. (to appear) and Math. Res. Center, Univ. of Wisconsin, Tech. Summary Report #1765, 1977.
9. G. Gripenberg, An existence result for a nonlinear Volterra integral equation in Hilbert space. SIAM J. Math. Anal. (to appear) and Helsinki Univ. of Tech. Report - HTKK-MAT-A86 (1976).

10. S.-O. Londen, On an integral equation in a Hilbert space. SIAM J. Math. Anal. (to appear) and Math. Res. Center, Univ. of Wisconsin, Tech. Summary Report #1527, 1975.

11. R. C. MacCamy, Stability theorems for a class of functional differential equations, SIAM J. Math. Anal. (to appear).

12. C. Rennolet, Abstract nonlinear Volterra integrodifferential equations of nonconvolution type, Ph.D. thesis, University of Wisconsin-Madison August 1977.

*Differential Equations and Applications*
*W. Eckhaus and E.M. de Jager (eds.)*
*©North-Holland Publishing Company (1978)*

# ON A NONLINEAR INTEGRAL EQUATION ARISING IN MATHEMATICAL EPIDEMIOLOGY

O. Diekmann
Mathematisch Centrum
2e Boerhaavestraat 49, Amsterdam

## 1. INTRODUCTION

In this note we shall study some qualitative aspects of the development of an epidemic in space and time. The mathematical problems that we shall come across are mainly those of proving existence or nonexistence of solutions of nonlinear convolution equations. We shall give a survey of some of the results that we obtained in [2] and, jointly with H.G. Kaper, in [3]; all details that we omit here can be found there.

In the space-independent Kermack and McKendrick model the evolution of the epidemic is governed by the equation

$$(1.1) \qquad u(t) = S_0 \int_0^t g(u(t-\tau))A(\tau)d\tau + \int_0^t h(\tau)d\tau, \qquad 0 \le t < \infty,$$

where

$$u(t) = -\ell n \frac{S(t)}{S_0}, \text{ S is the density of susceptibles,}$$

$$g(y) = 1 - e^{-y},$$

and A,h are given nonnegative functions describing, respectively, the infectivity of an individual which has been infected at $t = 0$ and the influence of the history up to $t = 0$. Suppose

$$\int_0^\infty A(\tau)d\tau = \gamma < \infty,$$

$$\int_0^\infty h(\tau)d\tau = H(\infty) < \infty,$$

then, under appropriate hypotheses, $u(t) \to u(\infty)$ as $t \to \infty$ and

$$(1.2) \qquad u(\infty) = \gamma S_0 g(u(\infty)) + H(\infty).$$

Equation (1.2) has a unique positive solution $u(\infty)$ for each positive $H(\infty)$. Let $\underline{u}$ be defined by

$$\underline{u} = \inf_{H(\infty)>0} u(\infty)$$

then clearly

$$(1.3) \qquad \underline{u} = \gamma S_0 g(\underline{u}).$$

The most important qualitative feature of the model is the so-called *treshold phenomenon*

$$\underline{u} > 0 \quad \text{if and only if} \quad \gamma S_0 > 1.$$

(Plotting a picture will make this evident.) The fact that this result is biologically significant appears from

$$\frac{S(\infty)}{S_0} < e^{-\underline{u}},$$

or, in words: the fraction of the susceptible population that escapes from getting the disease is less than $\exp(-\underline{u})$ for any initial infectivity (no matter how small).

Following Kendall we introduce space-dependence in the model by assuming that the infectivity is in fact a weighted spatial average

$$(1.4)\qquad u(t,x) = S_0 \int_0^t A(\tau) \int_{\mathbb{R}^n} g(u(t-\tau,\xi))V(x-\xi)d\xi d\tau + \int_0^t h(\tau,x)d\tau,$$

where $V: \mathbb{R}^n \to \mathbb{R}$ is a nonnegative radial function, and

$$\int_{\mathbb{R}^n} V(x)dx = 1.$$

An obvious question is now: is there an analogous threshold phenomenon for equation (1.4)? We shall show that the answer is yes if $n = 1$ or $n = 2$. In particular we find that, if $\gamma S_0 > 1$, the equation exhibits the *hair-trigger effect*: no matter how little infectivity is introduced in an arbitrarily small subset of $\mathbb{R}^n$, eventually there will be a large effect at every point.

Subsequently we shall investigate the possibility of *travelling wave solutions*. Instead of equation (1.4), describing an initial value problem, we then consider the time-translation invariant equation

$$(1.5)\qquad u(t,x) = S_0 \int_0^\infty A(\tau) \int_{\mathbb{R}^n} g(u(t-\tau,\xi))V(x-\xi)d\xi d\tau, \qquad -\infty < t < \infty.$$

Our main result is that, under appropriate hypotheses, there exists $c_0$, $0 < c_0 < \infty$, such that (1.5) has a travelling wave solution $u(t,x) = w(x+ct)$ if $|c| > c_0$ and no such solution if $|c| < c_0$.

In the following we shall normalize A and incorporate the constants $\gamma$ and $S_0$ in the function g.

## 2. THE HAIR-TRIGGER EFFECT

Let us consider the equation

$$(2.1)\qquad u(t,x) = \int_0^t A(\tau) \int_{\mathbb{R}^n} g(u(t-\tau,\xi))V(x-\xi)d\xi d\tau + f(t,x),$$

where $u: \mathbb{R}_+ \times \mathbb{R}^n \to \mathbb{R}$ is unknown and A, g, V and f satisfy

$H_A$: $A: \mathbb{R}_+ \to \mathbb{R}$ is nonnegative; $A \in L_1(\mathbb{R}_+)$ and $\int_0^\infty A(\tau)d\tau = 1$.

$H_g$: $g: \mathbb{R} \to \mathbb{R}$ is Lipschitz continuous (uniformly on $\mathbb{R}_+$), monotone nondecreasing and bounded from above; $g(0) = 0$.

$H_V$: $V: \mathbb{R}^n \to \mathbb{R}$ is nonnegative; $V \in L_1(\mathbb{R}^n)$ and $\int_{\mathbb{R}^n} V(x)dx = 1$; V is a radial function.

$H_f$: $f: \mathbb{R}_+ \times \mathbb{R}^n \to \mathbb{R}$ is nonnegative and continuous; $f(\cdot,x)$ is monotone nondecreasing for each $x \in \mathbb{R}^n$; $\{f(t,\cdot) \mid t \geq 0\}$ is uniformly bounded and equicontinuous.

Let $BC(\Omega)$ denote the Banach space of the bounded continuous functions on $\Omega$, equipped with the supremum norm.

THEOREM 2.1. *There exists a unique continuous solution* $u: \mathbb{R}_+ \times \mathbb{R}^n \to \mathbb{R}$ *of equation* (2.1). *Moreover* u *is nonnegative,* $u(\cdot,x)$ *is monotone nondecreasing for each* $x \in \mathbb{R}^n$ *and there exists* $u(\infty,\cdot) \in BC(\mathbb{R}^n)$ *satisfying*

$$(2.2)\qquad u(\infty,x) = \int_{\mathbb{R}^n} g(u(\infty,\xi))V(x-\xi)d\xi + f(\infty,x),$$

*and such that* $u(t,x) \to u(\infty,x)$ *as* $t \to \infty$ *uniformly on compact subsets of* $\mathbb{R}^n$.

SKETCH OF THE PROOF. The local (i.e., $t \in [0,T]$, $T$ sufficiently small) existence and uniqueness of a solution follows from a straightforward application of the Banach contraction mapping principle. The nonnegativity and the monotonicity follow from the construction of the solution as the limit of a sequence that is obtained by iteration. The global existence can be established by a continuation procedure and the boundedness by a simple estimate. The boundedness and the monotonicity yield pointwise convergence to a limit as $t \to \infty$ and by application of the Arzela-Ascoli theorem this can be strengthened as stated. Then it is easy to show that the limit has to satisfy equation (2.2). □

The mapping $N$ defined by $Nf = u$ relates the introduced infectivity to the thereby caused effect. Note that $Nf \equiv 0$ if $f \equiv 0$. For each finite $T$, $N$ is continuous as a mapping from $BC([0,T] \times \mathbb{R}^n)$ into itself. However, as a mapping from $BC(\mathbb{R}_+ \times \mathbb{R}^n)$ into itself, $N$ need not be continuous at $f \equiv 0$. In order to show this we shall investigate the final state equation (2.2). Firstly we state some lemmas which are crucial.

Let $w * k$ denote the convolution of $w$ and $k$ and $k^{m*}$ the $(m-1)$-times iterated convolution of $k$ with itself. Let $\hat{k}$ denote the Fourier transform of $k$. Suppose

$H_k$: $k \in L_1(\mathbb{R}^n)$, $k$ is nonnegative and $\int_{\mathbb{R}^n} k(x)dx = 1$.

Consider the inequality

$$(2.3) \qquad w \geq w * k.$$

LEMMA 2.2. *The following statements are equivalent*

(i) *there exists* $w \in BC(\mathbb{R}^n)$ *such that* (2.3) *is satisfied with strict inequality in some point;*

(ii) *there exists* $h > 0$ *such that*

$$\sum_{m=1}^{\infty} \int_{|x|\leq h} k^{m*}(x)dx$$

*converges;*

(iii) *there exist* $a > 0$, $C < \infty$ *such that*

$$\int_{|x|\leq a} \mathrm{Re}\left(\frac{1}{1+\varepsilon-\hat{k}(x)}\right) dx \leq C$$

*for all* $\varepsilon > 0$.

It is not easy to check directly whether a given $k$ satisfying $H_k$ has property (iii). Note that $\hat{k}(0) = 1$ and $\hat{k}(x) \neq 1$ for $x \neq 0$. So for $\varepsilon = 0$, the integrand in (iii) has a singularity at the origin and nowhere else. The next lemma establishes conditions on $k$ such that the singularity is integrable.

LEMMA 2.3.

(a) *If* $n = 1$ *and* $\int_{-\infty}^{\infty} |x|k(x)dx < \infty$ *then* (iii) *is equivalent to* $\int_{-\infty}^{\infty} xk(x)dx \neq 0$;

(b) *if* $n = 2$ *and* $\int_{\mathbb{R}^2} |x|^2 k(x)dx < \infty$ *then* (iii) *is equivalent to*

$$a \int_{\mathbb{R}^2} x_1 k(x)dx + b \int_{\mathbb{R}^2} x_2 k(x)dx \neq 0$$

*for some* $a$ *and* $b$;

(c) *if* $n \geq 3$ *then every* $k$ *satisfying* $H_k$ *has property* (iii).

A proof of lemmas 2.2 and 2.3 can be found in Essén [4] and in Feller [5, sections VI.10 and XVIII.7]. The following lemma deals with the case that the inequality (2.3) is in fact an equality.

LEMMA 2.4. *Suppose* $w \in BC(\mathbb{R}^n)$ *satisfies*

(2.4) $$w = w * k,$$

*then* $w \equiv C$ *for some constant* C.

For a proof see Rudin [6, Theorem 9.13] or Feller [5, section XI.2].

THEOREM 2.5. *Suppose that*
($\alpha$) *there exists* $p > 0$ *such that* $g(y) > y$ *for* $0 < y < p$ *and* $g(p) = p$,

($\beta$) $n = 1$ *or* $n = 2$ *and* $\int_{\mathbb{R}^n} |x|^n V(x)dx < \infty$,

($\gamma$) $f(\infty,\cdot)$ *is not identically zero,*

*then* $u(\infty,x) \geq p$ *for every* $x \in \mathbb{R}^n$.

PROOF. Let w be defined by $w(x) = \min\{u(\infty,x),p\}$, then (2.2) implies that w satisfies (2.3). Since V is a radial function, lemmas 2.2 and 2.3 imply that w has to satisfy (2.4) and hence that $w \equiv C$, $0 \leq C \leq p$. If $0 \leq C < p$ then (2.2) would not be satisfied. Hence $C = p$. □

In the special case that $g(y) = \gamma S_0(1-\exp(-y))$, the condition ($\alpha$) of Theorem 2.5 is equivalent to $\gamma S_0 > 1$. The lower bound p is independent of the function f. Thus we have demonstrated the hair-trigger effect if $\gamma S_0 > 1$.

## 3. TRAVELLING WAVES

Throughout the remaining part of the paper we shall assume that

$H_n$: $n = 1$.

$H_g^2$: g is two times continuously differentiable and there exists $p > 0$ such that $g(y) > y$ for $0 < y < p$ and $g(p) = p$.

$H_V^2$: V is continuous and has compact support.

Our results are actually valid under conditions on g and V which are weaker and which may be different for different theorems.

Substitution of $u(t,x) = w(x+ct)$ into equation (1.5) yields, upon some rearranging,

(3.1) $$w(\xi) = \int_{-\infty}^{\infty} g(w(\eta))V_c(\xi-\eta)d\eta, \quad \xi = x+ct,$$

where

(3.2) $$V_c(\xi) = \int_0^{\infty} A(\tau)V(\xi-c\tau)d\tau, \quad -\infty < \xi < \infty.$$

Since

$$\int_{-\infty}^{\infty} V_c(\xi)d\xi = 1,$$

(3.1) has, for every c, the constant solutions $w \equiv 0$ and $w \equiv p$. By a *nontrivial solution* of (3.1) we shall mean a continuous function w satisfying (3.1), $0 \leq w(x) \leq p$ and neither being identically 0 nor identically p. Theorem 2.5 shows that for $c = 0$ no nontrivial solutions do exist. As c increases, the mass of $V_c$ shifts to the right and in Theorem 3.1 we shall show that a nontrivial solution exists if the distribution of the mass of $V_c$ has become lopsided enough. Because of the symmetry of V we can restrict our attention to positive c.

THEOREM 3.1. *Suppose that* $g(y) \leq g'(0)y$ *for* $0 \leq y \leq p$, *then there exists* $c_0$, $0 < c_0 < \infty$, *such that for every* $c > c_0$ (3.1) *has a monotone nondecreasing solution* w *satisfying* $w(-\infty) = 0$, $w(\infty) = p$.

SKETCH OF THE PROOF. With the linearized equation

(3.3) $$v(\xi) = g'(0) \int_{-\infty}^{\infty} v(\eta)V_c(\xi-\eta)d\eta,$$

there is associated the characteristic equation

(3.4) $$L_c(\lambda) = 1,$$

where

(3.5) $$L_c(\lambda) = g'(0) \int_{-\infty}^{\infty} e^{-\lambda\xi} V_c(\xi)d\xi$$
$$= g'(0) \int_0^{\infty} e^{-\lambda c\tau} A(\tau)d\tau \int_{-\infty}^{\infty} e^{-\lambda\xi} V(\xi)d\xi.$$

Note that $L_c(0) = g'(0) > 1$ (since $y < g(y) \le g'(0)y$ for $0 < y \le p$). Real roots of (3.4) yield sign-definite solutions of (3.3). The constant $c_0$ is defined by

$$c_0 = \inf\{c \mid \text{there exists } \lambda > 0 \text{ such that } L_c(\lambda) = 1\}.$$

The nonnegativity of A and V guarantees that this definition makes sense (for fixed c, $L_c(\lambda)$ is a convex function of $\lambda$ and for fixed $\lambda$ it is a monotone decreasing function of c). The basic idea of the proof is to use information obtained from $L_c(\lambda)$ and the properties of g in the construction of two functions $\phi$ and $\psi$ such that $\phi \le \psi$, $T\phi \ge \phi$, and $T\psi \le \psi$, where T denotes the formal integral operator that is associated with the right-hand side of (3.1). The existence of a solution having the asserted properties is then established by means of an iterative process (T is monotone). □

Similar results have been obtained by Atkinson and Reuter [1] and by Weinberger [7]. Weinberger has also constructed functions $\phi$ and $\psi$ for the case $c = c_0$.

## 4. NONEXISTENCE AND UNIQUENESS

The first step towards a proof of the nonexistence of travelling waves with speed less than $c_0$ is provided by the following lemma concerning the convolution inequality (2.3).

LEMMA 4.1. *Suppose* k *satisfies* $H_k$ *for* $n = 1$ *and*

$$\int_{-\infty}^{\infty} |x|k(x)dx < \infty, \qquad \int_{-\infty}^{\infty} xk(x)dx \neq 0.$$

*Let* w *be a bounded and uniformly continuous solution of*

$$w \ge w * k.$$

*Then*

(i) $w - w*k \in L_1(\mathbb{R})$,

(ii) $\lim_{x\to-\infty} w(x)$ *and* $\lim_{x\to+\infty} w(x)$ *both exist,*

(iii) $$w(\infty) - w(-\infty) = \frac{\int_{-\infty}^{\infty} (w - w*k)(x)dx}{\int_{-\infty}^{\infty} xk(x)dx} .$$

SKETCH OF THE PROOF (see Essén [4] for a detailed proof).

Define

$$n(x) = \begin{cases} \int_x^{\infty} k(\xi)d\xi, & x > 0 \\ -\int_{-\infty}^{x} k(\xi)d\xi, & x \le 0, \end{cases}$$

then $n \in L_1(\mathbb{R})$ and

(4.1) $$w * n(x) - w * n(y) = \int_y^x (w - w*k)(\xi)d\xi.$$

From the monotonicity of the right-hand side and the boundedness of the left-hand side of (4.1) there follows (i) and

(4.2) $$w * n(\infty) - w * n(-\infty) = \int_{-\infty}^{\infty} (w - w*k)(\xi)d\xi.$$

Since

$$\hat{n}(\lambda) = \frac{1 - \hat{k}(\lambda)}{i\lambda} \text{ for } \lambda \neq 0$$

and

$$\hat{n}(0) = \int_{-\infty}^{\infty} xk(x)dx$$

we know that $\hat{n}(\lambda) \neq 0$ for all $\lambda$. Then (ii) and (iii) follow from (4.2) and Pitt's form of Wiener's general Tauberian Theorem (see for example Rudin [6, Theorem 9.7] or Widder [8, Theorem V.10a]. □

Assuming that

$H_A^2$: $\int_0^{\infty} \tau A(\tau)d\tau < \infty$

we have

COROLLARY 4.2. *Let* w *be a nontrivial solution of* (3.1) *for some* $c > 0$, *then*

$$\lim_{x\to-\infty} w(x) = 0 \quad \textit{and} \quad \lim_{x\to+\infty} w(x) = p.$$

PROOF. A bounded solution of (3.1) is necessarily uniformly continuous. Since

$$\int_{-\infty}^{\infty} xV_c(x)dx = c \int_0^{\infty} \tau A(\tau)d\tau > 0,$$

we deduce from Lemma 4.1 that $w(\infty) - w(-\infty) > 0$. From (3.1) and the properties of g it follows that only 0 and p are candidates for being limits. □

THEOREM 4.3. *Let the assumptions of Corollary* 4.2 *be satisfied and suppose* $g'(0) > 1$. *Then there exists* $a > 0$ *such that*

$$\int_{-\infty}^{\infty} w(x)\, e^{-\lambda x}\, dx$$

*converges for* $0 < \lambda < a$.

SKETCH OF THE PROOF. There exists $\ell > 1$ such that $g(w(x)) > \ell w(x)$ for $x \to -\infty$. Using this inequality and the same kind of arguments as those leading to (i) of Lemma 4.1, one can prove that $w \in L_1((-\infty,0))$ and subsequently by an induction process that

$$\int_{-\infty}^{0} |x|^m\, w(x)dx \leq m!\, a^{-m}$$

for some $a > 0$. □

Motivated by Theorem 4.3 we define

$$\bar{\lambda} = \sup\{\lambda \in \mathbb{R} \mid \int_{-\infty}^{\infty} w(x)\, e^{-\lambda x}\, dx \text{ converges}\}$$

and

$$W(\lambda) = \int_{-\infty}^{\infty} w(x)\, e^{-\lambda x}\, dx.$$

The function $W(\lambda)$ is analytic in the strip $0 < \text{Re}\,\lambda < \bar{\lambda}$. As a consequence of the nonnegativity of $w(x)$ we have (see Widder [8, Theorem II.5b])

LEMMA 4.4. *If* $\bar{\lambda} < \infty$, *then* $W(\lambda)$ *is singular in* $\lambda = \bar{\lambda}$.

Writing (3.1) as

$$w(x) = g'(0)\, w * V_c(x) + r(x),$$

where

$$r(x) = \int_{-\infty}^{\infty}\{g(w(\xi)) - g'(0)w(\xi)\}\, V_c(x-\xi)d\xi,$$

we obtain by Laplace transformation

(4.3) $\qquad W(\lambda) = W(\lambda)L_c(\lambda) + R(\lambda).$

If $\bar{\lambda} < \infty$ then Lemma 4.4 implies that $L_c(\lambda) = 1$ (note that $R(\lambda)$ is regular in a neighbourhood of $\lambda = \bar{\lambda}$). The possibility that $\bar{\lambda} = \infty$ can be excluded by a straightforward but technical proof. Thus we have established the following nonexistence result.

THEOREM 4.5. *Suppose* $c > 0$ *and* $L_c(\lambda) > 1$ *for* $\lambda \geq 0$, *then* (3.1) *has no nontrivial solution.*

Suppose, on the contrary, that the equation $L_c(\lambda) = 1$ has a positive real root, then (4.3) can be used to obtain information concerning the asymptotic behaviour, as $x \to -\infty$, of solutions of (3.1).

THEOREM 4.6. *Suppose* $c > c_0$ *and* $g(y) \leq g'(0)y$ *for* $0 \leq y \leq p$. *Let* $\sigma$ *denote the smallest positive root of* $L_c(\lambda) = 1$. *Let* $w$ *be a nontrivial monotone nondecreasing solution of* (3.1), *then there exists* $C > 0$ *such that*

$$\lim_{x\to-\infty} w(x)e^{-\sigma x} = C.$$

SKETCH OF THE PROOF. Note firstly that $L_c'(\sigma) \neq 0$ and $R(\sigma) \neq 0$. From (4.3) we obtain

$$W(\lambda) \sim \frac{R(\sigma)}{L_c'(\sigma)(\sigma-\lambda)}, \qquad \lambda \uparrow \sigma.$$

By a complex variable Tauberian theorem of the Ikehara type (see for instance Widder [8, Theorem V.17]) we can deduce from this formula the asymptotic behaviour of $w(x)$ as $x \to -\infty$. □

If $w(x)$ is a solution of (3.1), then so is every translate $w_{\bar{x}}(x) = w(x+\bar{x})$ of $w$. Our final theorem establishes a condition on $g$ such that every monotone nondecreasing nontrivial solution of (3.1) is obtained by translating one specific solution.

THEOREM 4.7. *Suppose* $c > c_0$ and

$$|g(y_1)-g(y_2)| \leq g'(0)|y_1-y_2| \quad \textit{for} \quad 0 \leq y_1,y_2 \leq p.$$

*Then there is modulo translation one and only one monotone nondecreasing nontrivial solution of* (3.1).

SKETCH OF THE PROOF. Let $w_1$ and $w_2$ be two monotone nondecreasing nontrivial solutions. By Theorem 4.6 we can find $\bar{x}$ such that $v$ defined by

$$v(x) = e^{-\sigma x}\{w_1(x) - w_2(x+\bar{x})\}$$

satisfies

$$\lim_{x\to-\infty} v(x) = \lim_{x\to+\infty} v(x) = 0.$$

From

$$|v(x)| \leq g'(0) \int_{-\infty}^{\infty} V_c(\xi)e^{-\sigma\xi}|v(x-\xi)|d\xi$$

one can deduce that $|v|$ cannot assume a maximum. Hence $v \equiv 0$. □

It is an open problem whether *every* nontrivial solution is monotone nondecreasing.

REFERENCES

[1] ATKINSON, C. & G.E.H. REUTER, *Deterministic epidemic waves*, Math. Proc. Camb. Phil. Soc. 80(1976) 315-330.

[2] DIEKMANN, O., *Thresholds and travelling waves for the geographical spread of infection*, Mathematical Centre Report TW 166/77, Amsterdam, 1977.

[3] DIEKMANN, O. & H.G. KAPER, *On the bounded solutions of a nonlinear convolution equation*, in preparation.

[4] ESSÉN, M., *Studies on a convolution inequality*, Ark. Mat. 5(1963) 113-152.

[5] FELLER, W., *An Introduction to Probability Theory and Its Applications*, Vol. II (Wiley, New York, 1966).
[6] RUDIN, W., *Functional Analysis* (McGraw-Hill, New York, 1973).
[7] WEINBERGER, H.F., *Asymptotic behavior of a model in population genetics*, to appear in: J. Chadam, ed., Indiana University Seminar in Applied Mathematics, Springer Lecture Notes.
[8] WIDDER, D.V., *The Laplace Transform* (University Press, Princeton, 1946).

REMARK. We also draw attention to the recent paper

D.G. ARONSON, *The asymptotic speed of propagation of a simple epidemic*, to appear in: W.E. Fitzgibbon & H.F. Walker, eds., Nonlinear Diffusion (Research Notes in Mathematics, Pitman Publishing Co., 1977).

There it is shown that for a special case of the model $c_0$ is the asymptotic speed of propagation of infection.

REMARK. In the final draft of the paper [3] an alternate proof of Theorem 4.6. is given without presupposing that the solution be monotone. Hence, the monotonicity condition in Theorem 4.7 can be dispensed with. We observe that this leads, in a very indirect way, to the conclusion that indeed every nontrivial solution is monotone nondecreasing. Still it would be of interest to have a direct proof of this fact.

*Differential Equations and Applications*
*W. Eckhaus and E.M. de Jager (eds.)*
*©North-Holland Publishing Company (1978)*

# DOWNSTREAM DEVELOPMENT OF VELOCITY-PROFILES BEHIND FLAT PLATES

C.J. van Duyn

Delft University of Technology,*

Delft, Netherlands

## 1. Introduction

In this paper we shall discuss a classical problem from laminar boundary layer theory, namely that of the downstream development of velocity profiles behind flat plates. We shall consider the two dimensional case, in which a flat plate of length L is placed in a steady, laminar flow of an incompressible fluid, and we choose the coordinates so that the plate is situated at $y = 0$, $-L \leq x \leq 0$.

About the velocity of the flow we shall assume that for large values of $|y|$ it is constant and parallel to the plate, such that for $y \to \infty$ it is $U_1$ and for $y \to -\infty$ it is $U_2$, where $U_1$ and $U_2$ are both positive constants.

Then, if $U_1 = U_2$ we have the classical wake problem for a flat plate, and if $U_1 \neq U_2$ we have behind the plate two laminar streams which move at different velocities and which interact through friction.

In the case where $U_1 = U_2$ we shall show that the velocity converges towards $U_1$ at large distances downstream, and in the case where $U_1 \neq U_2$ the velocity profile will converge upon an appropriately chosen similarity profile.

In both cases, we shall obtain for the rate of convergence

---

*At present at Leiden University, Mathematical Institute, Leiden, Netherlands.

$O(x^{-\frac{1}{2}(1-\varepsilon)})$, in which $\varepsilon$ may be any positive number, and we shall show that this convergence result hardly depends on the velocity profile at the trailing edge of the plate.

Let u, v be the velocity components in the x, y direction, and let Q denote the domain $0 < x < \infty$, $-\infty < y < \infty$. Then according to laminar boundary layer theory (cf.SCHLICHTING [10], MEYER [4]), this problem is described by the Prandtl equations for a two-dimensional steady laminar flow

$$u_x + v_y = 0 \quad ,$$

and

$$u\, u_x + v\, u_y = \nu u_{yy} \, ,$$

in Q. Here $\nu$ is the kinematical viscosity, which is a positive constant.

The boundary conditions are

$$u(0,y) = u_0(y) \quad \text{for} \quad -\infty < y < \infty \, ,$$

where $u_0$ is a given initial profile, together with Prandtl's matching - conditions

$$u(x,y) \to U_1 \quad \text{when} \quad y \to \infty \, ,$$

and

$$u(x,y) \to U_2 \quad \text{when} \quad y \to -\infty ,$$

in which the convergence is pointwise in x.

Further, we have the additional condition

$$v(x,0) = 0 \quad \text{for all} \quad 0 < x < \infty \, .$$

In what follows, we shall refer to this problem as Problem I.

About the initial profile $u_0$ we shall assume that it satisfies

(i) $u_0(0) = 0$ , $u_0(y) > 0$ when $y \in \mathbb{R}\setminus\{0\}$;

(ii) there exist constants $K_1$, $K_2 \in (0,\infty)$ such that $K_1|y| \le u_0(y) \le K_2|y|$ for $|y| \le y_1$, where $y_1 > 0$ ;

(iii) $u_0(y) \to U_1$ as $y \to \infty$ and $u_0(y) \to U_2$ as $y \to -\infty$.

These assumptions are natural to the problem, because the velocity profile at the trailing edge of the plate has been strongly influenced by the plate, and therefore it will have a boundary layer like profile.

In section 2 we shall demonstrate that if $u_0$ is uniformly Lipschitz continuous on $\mathbb{R}$ such that it satisfies assumptions (i) - (iii), then Problem I has a unique classical solution $(u,v)$, such that $u > 0$ on $Q$.

In order to deal with the question of asymptotic behaviour and to use some earlier results on this subject (VAN DUYN & PELETIER [2]), we change to von Mises variables, what will cause a considerable simplification of Problem I. More precisely, this transformation reduces Problem I to a Cauchy problem for a nonlinear diffusion equation [4],[10].

Let

$$x = x \ , \ \psi = \psi(x,y) = \int_0^y u(x,s)ds - \int_0^x v(t,0)dt \ .$$

Because $u > 0$ on $\bar{Q}\backslash\{(0,0)\}$ and satisfies the matching-conditions, and because $v(x,0) = 0$ for all $x\in(0,\infty)$, this is a one-to-one map from $\bar{Q}$ to the region $0 \leq x < \infty$, $-\infty < \psi < \infty$.

When we now regard $u$ as a function of $x$ and $\psi$, Problem I becomes

$$u_x = \tfrac{1}{2}\nu(u^2)_{\psi\psi} \text{ in } S = \{(x,\psi) : 0 < x < \infty, -\infty < \psi < \infty\}, \quad (1)$$

$$u(0,\psi) = u_0(\psi) \quad \text{for} \quad \psi\in \mathbb{R}, \quad (2)$$

$$u(x,\psi) \to U_1 \quad \text{when } \psi \to +\infty, \quad (3)$$

$$u(x,\psi) \to U_2 \quad \text{when } \psi \to -\infty, \quad (4)$$

where the convergence is pointwise in $x$.

We shall call this Problem II.

Equation (1) can be transformed into an ordinary differential equation by introducing the similarity variable $\eta = \psi(x + 1)^{-\frac{1}{2}}$. The corresponding similarity solution $u(x,\psi) = f(\eta)$ should then satisfy

$$\tfrac{1}{2}\nu(f^2)'' + \tfrac{1}{2}\eta f' = 0 \quad \text{on } \mathbb{R}. \quad (5)$$

Here, a prime denotes differentiation with respect to $\eta$.

In order to compare the solutions of (5) with those of Problem II, we shall require at the boundaries

$$f(-\infty) = A \ , \quad f(\infty) = B \ . \tag{6}$$

Recently, it was shown by VAN DUYN & PELETIER [3] that problem (5),(6) has a unique solution for any $A \geq 0$ and for any $B \geq 0$, such that at points where $f > 0$, the solution is strictly monotonically increasing if $A < B$ and strictly monotonically decreasing if $A > B$. Further, since equation (5) degenerates at those points where $f = 0$, they showed that we can distinguish the following cases:

1. $A > 0$ , $B > 0$.

Then $f \in C^{\infty}(\mathbb{R})$ and, assuming $A < B$, $f(\eta) \in (A,B)$ for all $\eta \in \mathbb{R}$.

2. $A = 0$ , $B > 0$.

Then $f \in C(\mathbb{R})$ and there exists a negative number b, which depends on B, such that

$$f(\eta) = 0 \text{ for } \eta \in (-\infty, b],$$

and

$$f(\eta) \in (0,B) \text{ for } \eta \in (b,\infty).$$

3. $A > 0$ , $B = 0$.

Then $f \in C(\mathbb{R})$ and there exists a positive number a, which depends on A, such that

$$f(\eta) \in (0,A) \text{ for } \eta \in (-\infty, a),$$

and

$$f(\eta) = 0 \quad \text{for } \eta \in [a,\infty).$$

Moreover, PELETIER [8] showed that whenever $f(\eta) \to C$ as $|\eta| \to \infty$, with $C > 0$, then f satisfies the asymptotic behaviour

$$f(\eta) - C = O(\operatorname{erfc}\,[|\eta|/2(\nu C)^{\frac{1}{2}}]) \text{ as } |\eta| \to \infty. \tag{7}$$

In general, a solution of problem (5),(6) will be denoted by $f(\eta;A,B)$ and for convenience we shall often replace $f(\eta;U_2,U_1)$ simply by $f(\eta)$.

Besides establishing existence and uniqueness of a classical solution of Problem I, the main purpose of this paper is to prove convergence of a solution of Problem I towards the similarity profile $f(\eta)$ as $x \to \infty$.

Before we can give a precise statement, we shall have to impose a restriction on the convergence of $u_0(y)$ as $|y|$ tends to infinity.

Let $M = \sup_{\mathbb{R}} u_0(y)$. Then we shall require that

A1. $u_0(y) - U_1 = O(\operatorname{erfc}[\, My/2(\nu U_1)^{\frac{1}{2}}])$ as $y \to \infty$.

A2. $u_0(y) - U_2 = O(\operatorname{erfc}[-My/2(\nu U_2)^{\frac{1}{2}}])$ as $y \to -\infty$.

These assumptions imply that the transformed initial value $u_0(\psi)$ satisfies the same asymptotic behaviour as $f$ in (7), where $C = U_1$ as $\psi \to \infty$ and $C = U_2$ as $\psi \to -\infty$.

Then following the method developed in [2], we shall obtain

THEOREM 1. _Let_ $u(x,y)$ _be a solution of Problem I, in which_ $u_0(y)$ _satisfies the additional conditions_ A1 _and_ A2. _Let_ $\bar{u}(x,y)$ _be the corresponding similarity profile of Problem I, considered as a function of the physical variables_ $x$ _and_ $y$. _Then for each_ $\varepsilon > 0$, _there exist constants_ $K(\varepsilon)$ _and_ $K_0$ _such that for all_ $(x,y) \in \bar{Q}$

$$|u(x,y) - \bar{u}(x,y)| \leq K(\varepsilon)(x+1)^{-\frac{1}{2}(1-\varepsilon)} . e^{K_0 . \Theta(x,y)},$$

_where_

$$\Theta(x,y) = y.(x+1)^{-\frac{1}{2}}.$$

So, if we move in the positive x-direction along parabolas of the form $y = \text{const.}(x+1)^{\frac{1}{2}}$, then the velocity component in the x-direction -$u(x,y)$- will converge towards the similarity profile -$\bar{u}(x,y)$- and the rate of convergence will be arbitrary close to $O(x^{-\frac{1}{2}})$. The parabolas $y = \text{const.}(x+1)^{\frac{1}{2}}$ are natural for this problem. It is well known that the boundary layer equations have similarity solutions of the form $f(\Theta)$, where $\Theta = y(\frac{\nu}{U_1}(x+1))^{-\frac{1}{2}}$ and $f$ satisfies the Blasius equation $2f''' + f\,f'' = 0$ (cf. SCHLICHTING [10], p. 175).

Acknowledgement

The author is greatly indebted to Professor L.A. Peletier for the many stimulating discussions on this subject and for reading the manuscript.

## 2. Existence and uniqueness of a classical solution

In this section we shall prove the existence and uniqueness of a classical solution of Problem I. More precisely, we shall show that given a suitably chosen initial value $u_0(y)$, there exists a unique solution $(u,v)$ of Problem I such that $u \in C(\bar{Q})$ and $u_x, u_y, u_{yy}, v, v_y \in C(Q)$.

We shall follow a method which is based on the inverse von Mises transformation. This idea has also been used by OLEINIK [5], who proved existence and uniqueness of a classical solution of the boundary layer equations, which then described the flow of an incompressible fluid along a flat plate: that is, she studied Problem I in the domain $x > 0$, $y > 0$ with the additional boundary condition $u(x,0) = 0$.

Let the initial value $u_0(y)$ be uniformly Lipschitz continuous on $\mathbb{R}$ such that it satisfies the assumptions (i) - (iii) and let

$$\Psi(y) = \int_0^y u_0(t)dt \ .$$

Then it is clear that $\Psi$ is a one-to-one map from $\mathbb{R}$ onto $\mathbb{R}$.

As the first step of the proof, we consider a function $w_0(s)$ on $\mathbb{R}$, which is uniquely defined by the expression

$$w_0(s) = u_0^2(\Psi^{-1}(s)) \ , \text{ for } s \in \mathbb{R} \ . \tag{8}$$

Because of the assumptions on $u_0$, it is easy to see that $w_0(s)$ is uniformly Lipschitz continuous on $\mathbb{R}$ and satisfies

(i') $w_0(0) = 0$ , $w_0(s) > 0$ for $s \in \mathbb{R}\backslash\{0\}$;

(ii') $c_1|s| \leq w_0(s) \leq c_2(s)$ for $|s| < s_1$, where $c_1, c_2 \in (0,\infty)$ and

$$s_1 = \int_0^{y_1} u_0(t)dt \ ;$$

(iii') $w_0(s) \to U_1^2$ as $s \to +\infty$ and $w_0(s) \to U_2^2$ as $s \to -\infty$.

Then, if $S_X$ denotes the strip $\{(x,s) : x \in (0,X], s \in \mathbb{R}\}$, where X is some fixed positive number, we consider the initial value problem

$$\text{I}'\quad \begin{cases} w_x = \nu w^{\frac{1}{2}} w_{ss} & \text{in } S_X , \\ w(0,s) = w_0(s) & \text{on } \mathbb{R} . \end{cases} \qquad (9)$$

Equation (9) is degenerate parabolic: i.e. at points where $w > 0$ the equation is parabolic, while at points where $w = 0$ it is not. Therefore, we have to interpret the solutions of this problem in some weak sense. Applying the results of OLEINIK, KALASHNIKOV and YUI-LIN [7], who defined a suitable class of generalized solutions, it follows that Problem I' has a unique solution within this class, which is continuous and bounded on $\bar{S}_X$ and which satisfies the equation in a classical sense, in a neighbourhood of those points where $w$ is positive. Moreover they showed that the weak maximum principle is valid for its solutions.

First we shall prove that a generalized solution of Problem I', in which $w_0$ satisfies (i') - (iii'), is positive on $S_X$. We shall make this the content of the next lemma.

LEMMA 1. *Let $w$ be a generalized solution of Problem I', in which $w_0$ satisfies (i') - (iii'). Then for each $x_p \in (0,X]$, there exists a positive constant $\Delta(x_p)$ such that $w(x,s) \geq \Delta(x_p)$ for all $x \in [x_p, X]$ and for all $s \in \mathbb{R}$.*

In what follows, we shall frequently use two different kinds of similarity solutions of equation (9). Let $\eta = s(x+1)^{-\frac{1}{2}}$ denote the similarity variable. Then we shall consider similarity solutions of the form

1. $f^2(\eta;A,B)$, $-\infty < \eta < \infty$, $A \geq 0$, $B \geq 0$.

Similarity solutions of this type arise, because $w^{\frac{1}{2}}$ satisfies equation (1) in which the independent variable $\psi$ has been replaced by $s$. Therefore it is clear that the square of solutions of (5) can serve as similarity solutions of equation (9).

2. $f^2(\eta;A)$ , $0 < \eta < \infty$, $A \geq 0$.

The function $f(\eta;A)$ satisfies equation (5) and the boundary conditions

$f(0) = 0$ and $f(\eta) \to A$ as $\eta \to \infty$. Moreover, it can be shown that the derivative $\{f^2\}'$ at $\eta = 0$ satisfies $\{f^2(0;A)\}' = \text{const}.A^{3/2}$ (cf. PELETIER [9], further references are given there).

PROOF OF LEMMA 1. CRAVEN & PELETIER [1] showed that the derivative with respect to $\eta$ of a similarity solution of the type $f(\eta;A,0)$, with $\lim_{\eta \to a^-} f(\eta;A,0) = 0$ satisfies $\lim_{\eta \to a^-} f'(\eta;A,0) = -\frac{1}{2}\frac{a}{\nu}$.
Therefore, if we take $A < \min\{U_1,U_2\}$ and sufficiently small we obtain for all $s \in \mathbb{R}$

$$w_0(s) \geq f^2(s + a; A,0) ,$$

and

$$w_0(s) \geq f^2(s - a; 0,A) .$$

Then using the maximum principle, we find

$$w(x,s) \geq \max\{f^2((s+a)(x+1)^{-\frac{1}{2}};A,0);f^2((s-a)(x+1)^{-\frac{1}{2}};0,A)\} ,$$

for all $(x,s) \in \bar{S}_X$.
Next we observe that a similarity solution of the form $f((s+a)(x+1)^{-\frac{1}{2}};A,0)$ has a front in the x-s plane which moves along the parabola $s=a(x+1)^{\frac{1}{2}}-a$, and $f > 0$ in the region $x \geq 0$, $s < a(x+1)^{\frac{1}{2}}-a$ and $f = 0$ for $x \geq 0$, $s \geq a(x+1)^{\frac{1}{2}}-a$. A similar behaviour occurs for solutions of the form $f((s-a)(x+1)^{-\frac{1}{2}};0,A)$, where the parabola is now given by $s = -a(x+1)^{\frac{1}{2}}+a$. Then, if we use these observations in the lower bound on w, we obtain the desired result.

As mentioned before, this positivity of w on $S_X$ implies that Problem I' has a unique classical solution: i.e. $w \in C^\infty(S_X) \cap C(\bar{S}_X)$. Further, by using a similar argument as in the proof of Lemma 2, it can be shown that for each $x \in [0,X]$, $w(x,s) \to U_1^2$ as $s \to \infty$ and $w(x,s) \to U_2^2$ as $s \to -\infty$.

Let $Q_X = \{(x,y) : x \in (0,X], s \in \mathbb{R}\}$. Then, our next step is to define a function $\psi(x,y)$ on $\bar{Q}_X$ through the expression

$$y = \int_0^{\psi(x,y)} (w(x,s))^{-\frac{1}{2}} ds \ . \tag{10}$$

PROPOSITION 1. *The function* $\psi(x,y)$ *is well defined on* $\bar{Q}_X$.

PROOF. Define the function $g(s;A)$ on $\mathbb{R}$ by

$$g(s;A) = \begin{cases} f^2(s;A) & \text{for } s \geq 0 \ , \\ f^2(-s;A) & \text{for } s \leq 0. \end{cases}$$

Now, using the fact that $f^2(s;A)$ is a concave function and regarding its behaviour as $s \to 0$ and $s \to \infty$, it is clear that a constant A can be chosen so that $w_0(s) \geq g(s;A)$ for all $s \in \mathbb{R}$. Then following a maximum principle argument (PELETIER [9], p. 114; SERRIN [11], Lemma 1) we find that $w(x,s) \geq g(s(x+1)^{-\frac{1}{2}};A)$ for all $(x,s) \in \bar{S}_X$. Moreover, using again the behaviour of $f^2(\eta;A)$ as $\eta \to 0$, we obtain that for fixed $x \in [0,X]$ the function $\{g(s(x+1)^{-\frac{1}{2}};A)\}^{-\frac{1}{2}}$ is integrable with respect to s on bounded intervals. So for each $x \in [0,X]$ and each $y \in \mathbb{R}$, there exists a unique number $\psi(x,y)$ such that $y = \int_0^{\psi(x,y)} (w(x,s))^{-\frac{1}{2}} ds$. Hence the function $\psi(x,y)$ is well-defined on $\bar{Q}_X$.

Now we turn to the matter of continuity.

PROPOSITION 2. $\psi(x,y) \in C(\bar{Q}_X)$.

PROOF. Let $(x_0,y_0)$ be a fixed but arbitrary point in $\bar{Q}_X$. Then for any $(x,y) \in \bar{Q}_X$, we can write (10) in the form

$$\int_{\psi(x_0,y_0)}^{\psi(x,y)} (w(x,s))^{-\frac{1}{2}} ds = y - y_0 + \int_0^{\psi(x_0,y_0)} \{(w(x_0,s))^{-\frac{1}{2}} - (w(x,s))^{-\frac{1}{2}}\} ds.$$

Because a solution of Problem I' is bounded on $\bar{S}_X$, there exists a constant $M_0 < \infty$ such that

$$M_0^{-\frac{1}{2}} |\psi(x,y) - \psi(x_0,y_0)| \leq |y - y_0| + \left| \int_0^{\psi(x_0,y_0)} \{(w(x_0,s))^{-\frac{1}{2}} - (w(x,s))^{-\frac{1}{2}}\} ds \right| . \tag{11}$$

Now observe that for each $s\in \mathbb{R}\backslash\{0\}$, $(w(x,s))^{-\frac{1}{2}} \to (w(x_0,s))^{-\frac{1}{2}}$ as $x \to x_0$, and that $(w(x,s))^{-\frac{1}{2}} \leq \{g(s(X+1)^{-\frac{1}{2}};A)\}^{-\frac{1}{2}}$ for $s\in \mathbb{R}\backslash\{0\}$, $0 \leq x \leq X$. Hence, applying the Dominated Convergence Theorem to the second term in the right hand side of (11), we obtain that $\psi(x,y) \to \psi(x_0,y_0)$ when $(x,y) \to (x_0,y_0)$.

Now we can carry out the final step and execute the inverse von Mises transformation. Define the functions

$$u(x,y) = (w(x,\psi(x,y)))^{\frac{1}{2}} \text{ on } \bar{Q}_X,$$

and

$$v(x,y)=\tfrac{1}{2}(w(x,\psi(x,y)))^{\frac{1}{2}}.\int_0^{\psi(x,y)} \{w_x(x,s).w^{-3/2}(x,s)\}ds \text{ on } Q_X.$$

From Propositions 1 and 2, and the continuity of $w$ it follows that $u\in C(\bar{Q}_X)$, and from the positivity and smoothness of $w$ in $S_X$, one can easily establish in a straightforward manner that $v,v_y,u_x,u_y,u_{yy}\in C(Q_X)$ and satisfy

$$u_x + v_y = 0 ,$$

and

$$u\, u_x + v\, u_y = \nu u_{yy} \quad \text{in } Q_X .$$

Moreover, for each $x\in[0,X]$, $u(x,y) \to U_1$ as $y \to \infty$ and $u(x,y) \to U_2$ as $y \to -\infty$. Also, since $\psi(x,0) = 0$ on $[0,X]$, we find that $v(x,0) = 0$ on $(0,X]$. Finally, since $u\in C(\bar{Q}_X)$ and has been uniquely defined by $w$, it follows that $u(x,y) \to u_0(y)$ as $x \to 0^+$ for all $y\in \mathbb{R}$. Now, remembering that $X$ had been chosen arbitrary, we have proved the following theorem.

THEOREM 2. *Let $u_0$ be uniformly Lipschitz continuous, such that it satisfies assumptions (i) - (iii). Then Problem I has a unique classical solution in the domain $0 \leq x < \infty$, $-\infty < y < \infty$.*

## 3. Integral estimate

In this section we shall first derive a preliminary bound on the solution $u$ of Problem II. Then we shall use this bound to show that a solution of Problem II converges, as $x \to \infty$, towards the corresponding similarity solution $f(\eta)$, and we obtain an integral estimate for the rate of convergence. In section 5 we shall use this integral estimate to construct a pointwise estimate for the rate of convergence, which will yield Theorem 1 after a transformation to the original variables x and y.

We pointed out in the introduction, that if $u_0(y)$ satisfies the assumptions A1 and A2, then $u_0(\psi)$ has the same asymptotic behaviour as $f(\psi)$ when $|\psi| \to \infty$. So $u_0(\psi)$ satisfies:

A1' $\qquad u_0(\psi) - U_1 = O(\operatorname{erfc}[\psi/2(\nu U_1)^{\frac{1}{2}}]) \quad \text{as } \psi \to \infty,$

and

A2' $\qquad u_0(\psi) - U_2 = O(\operatorname{erfc}[-\psi/2(\nu U_2)^{\frac{1}{2}}]) \quad \text{as } \psi \to -\infty.$

LEMMA 2. *Let* $u(x,\psi)$ *be the solution of Problem II, in which* $u_0(\psi)$ *satisfies A1' and A2'. Then there exist numbers* $U_1^+ \geq U_1$, $U_2^+ \geq U_2$, $\gamma_1 \leq u_1 < 0$ *and* $\gamma_2 \geq u_2 > 0$, *with* $f(u_1;\, 0,U_1) = 0$ *and* $f(u_2;\, U_2,0) = 0$ *such that*

$$\max\{f((\psi+\gamma_2)(x+1)^{-\frac{1}{2}};U_2,0),f((\psi+\gamma_1)(x+1)^{-\frac{1}{2}};0,U_1)\} \leq u(x,\psi) \leq$$

$$\min\{f(\psi(x+1)^{-\frac{1}{2}};\, U_2,U_1^+),f(\psi(x+1)^{-\frac{1}{2}};\, U_1,U_2^+)\},$$

*for all* $(x,\psi)\in\bar{S}$.

PROOF. We know that $f(\psi;U_2,0)$ satisfies A2'. Hence, shifting this function to the left, it is clear that there exists a number $\gamma_2 \geq u_2 > 0$ such that $u_0(\psi) \geq f(\psi+\gamma_2;U_2,0)$ for all $\psi\in\mathbb{R}$. In the same way we can see that there exists a number $U_1^+ \geq U_1$ such that $u_0(\psi) \leq f(\psi;U_2,U_1^+)$ for all $\psi\in\mathbb{R}$. Hence, by the maximum principle

$$f((\psi+\gamma_2)(x+1)^{-\frac{1}{2}};U_2,0) \leq u(x,\psi) \leq f(\psi(x+1)^{-\frac{1}{2}};U_2,U_1^+),$$

for all $(x,\psi)\in\bar{S}$.

The second half of the proof can be found in an identical manner and we shall omit further details.

We now turn to the question of convergence. In order to compare a solution $u = u(x,\psi)$ of Problem II with a similarity solution, which only depends on $\eta$, we transform to the new independent variables $\eta$ and $\xi = \log(1+x)$. Then, with $u$ regarded as a function of $\eta$ and $\xi$, equation (1) becomes

$$u_\xi = \tfrac{1}{2}\nu(u^2)_{\eta\eta} + \tfrac{1}{2}\eta u_\eta, \tag{12}$$

in the halfspace $(-\infty,\infty) \times (0,\infty)$, which we shall denote again by $S$. At $x = 0$, we have $\xi = 0$ and $\eta = \psi$. So

$$u(\eta,0) = u_0(\eta) \quad \text{for all } \eta\in(-\infty,\infty) \tag{13}$$

Next we define the function

$$\Phi(\xi) = \int_{-\infty}^{+\infty} \{u(\xi,\eta) - f(\eta)\}d\eta$$

By Lemma 2, $\Phi(\xi)$ is well defined for all $\xi \geq 0$.

LEMMA 3. *The function* $\Phi(\xi)$ *satisfies*

$$\Phi(\xi) = \Phi(\delta)\, e^{-\frac{1}{2}(\xi-\delta)} \quad \textit{for all } \xi \geq \delta,$$

*where* $\delta$ *may be any positive constant. Moreover if* $u_0'(\psi) \to 0$ *as* $|\psi| \to \infty$, *we may set* $\delta = 0$.

PROOF. Because the similarity solution only depends on $\eta$, we may write equation (5) as $f_\xi = \frac{1}{2}\nu(f^2)_{\eta\eta} + \frac{1}{2}\eta f_\eta$. Then, if we subtract this equation from equation (12) and integrate with respect to $\eta$ from $-\infty$ to $+\infty$, we obtain

$$\frac{d}{d\xi}\Phi(\xi) = \tfrac{1}{2}\nu(u^2-f^2)_\eta\Big|_{-\infty}^{+\infty} + \tfrac{1}{2}\eta(u-f)\Big|_{-\infty}^{+\infty} - \tfrac{1}{2}\Phi(\xi).$$

By Lemma 2, $(u-f)$ tends to zero as $|\eta| \to \infty$ sufficiently fast, so that $\eta(u-f) \to 0$ as $|\eta| \to \infty$. Also $f_\eta \to 0$ as $|\eta| \to \infty$. Next, using Lemma 2 again, we notice that equation (12) is uniformly parabolic for large values of $|\eta|$ and all $\xi \geq 0$. Then anticipating the proof of Lemma 4, we find: $u_\eta(\xi,\eta) \to 0$ as $|\eta| \to \infty$ for all $\xi \geq \delta$, where $\delta$ may be any positive constant. Moreover, if $u_0'(\psi) \to 0$ as $|\psi| \to \infty$ we may set $\delta = 0$. Therefore

$$\frac{d}{d\xi}\Phi(\xi) = -\tfrac{1}{2}\Phi(\xi) \quad \text{for all } \xi \geq \delta,$$

and Lemma3 follows by integrating this expression.

Now define

$$u_0^+(\eta) \geq \max\{u_0(\eta), f(\eta)\}\ , \qquad (14)$$

and

$$u_0^-(\eta) \leq \min\{u_0(\eta), f(\eta)\}\ , \qquad (15)$$

which have the same properties as $u_0(\eta)$ and in addition are chosen so that $(u_0^+)'$ and $(u_0^-)'$ tend to zero as $|\eta| \to \infty$. Denote the solutions of (12), (14) and (12), (15) by $u^+(\xi,\eta)$ and $u^-(\xi,\eta)$, respectively. Then if we apply Lemma 3 to $(u^+ - f)$ and $(u^- - f)$ and if we use the weak maximum principle, we obtain

$$\int_{-\infty}^{+\infty} |u(\xi,\eta) - f(\eta)|d\eta \leq \int_{-\infty}^{+\infty} \{u_0^+(\eta) - u_0^-(\eta)\}d\eta . e^{-\frac{1}{2}\xi}\ ,$$

for all $\xi \geq 0$.

Hence we proved the following convergence theorem for solutions of Problem II.

THEOREM 3. *Let* $u(x,\psi)$ *be the solution of* Problem II, *in which* $u_0(\psi)$ *satisfies* A1' *and* A2', *and let* $\tilde{u}(x,\eta) \equiv u(x,\psi)$. *Then there exists a constant* K, *depending only on the data of the problem, such that*

$$\int_{-\infty}^{+\infty}|\tilde{u}(x,\eta) - f(\eta)|d\eta \leq K(x+1)^{-\frac{1}{2}} \text{ for all } x \geq 0.$$

## 4. A regularity property

In order to convert the integral estimate (Theorem 3) into a pointwise estimate, we need a regularity property of the solutions u and f. More precisely, we shall need an estimate for the derivative $(u^2 - f^2)$ on lines of constant $\xi_0$ in the $\eta - \xi$ plane, in terms of the supremum of $(u^2 - f^2)$ taken over lines of constant $\xi_0 - \delta$, where $\delta$ is a fixed but arbitrary positive constant. We shall derive such an estimate by following the method developed in [2], which is basically an application of the Bernstein argument to uniformly parabolic differentual equations.

Let $\xi_p = \log(1+p)$, where p may be any positive number, and denote the region $\xi > \xi_p$, $-\infty < \eta < \infty$ by $S^p$. Then it follows from Lemma 1 that $u > \Delta^{\frac{1}{2}}(p)$ on $S^p$, and hence equation (12) is uniformly parabolic in $S^p$. From now on we shall restrict the analysis in this section to $S^p$.

First we shall prove that $|u_\eta|$ is uniformly bounded on $S^{p+\delta}$, with $\delta > 0$.

LEMMA 4. *Let $u(\xi,\eta)$ be a solution of (12), (13), in which $u_0$ satisfies A1' and A2'. Let $\delta > 0$. Then there exists a constant $C_1$ such that*

$$|u_\eta(\xi,\eta)| \leq C_1 \text{ for all } \eta \in \mathbb{R} \text{ and } \xi \geq \xi_p + \delta.$$

PROOF. Let $R(\rho,\sigma)$ denote the rectangle $(\rho-1,\rho+1) \times (\sigma,\sigma+1]$ and let $R_\delta(\rho,\sigma)$ be the rectangle $(\rho-1-\delta,\rho+1+\delta) \times (\sigma-\delta,\sigma+1]$, where $\rho \in \mathbb{R}$ and $\sigma \geq \xi_p + \delta$. Let

$$\underset{R_\delta}{\operatorname{osc}}\, u = \sup_{R_\delta} u - \inf_{R_\delta} u .$$

Then, using a Bernstein-technique, it can be shown (cf. OLEINIK & KRUZHKOV [6]) that

$$\sup_{R(\rho,\sigma)} |u_\eta| \leq C(1+|\rho|)^{\frac{1}{2}} \underset{R_\delta(\rho,\sigma)}{\operatorname{osc}}\, u \tag{16}$$

where the constant C does not depend on $\rho$ and $\sigma$. Now from Lemma 2 and the asymptotic behaviour of the similarity solutions it follows that for any $r \in \mathbb{R}$

$$|\rho|^r \underset{R_\delta(\rho,\sigma)}{\operatorname{osc}}\, u \to 0 \quad \text{as } |\rho| \to \infty,$$

uniformly with respect to $\sigma \geq \xi_p + \delta$. Therefore the right-hand-side of (16) is uniformly bounded for $\rho \in \mathbb{R}$ and $\sigma \geq \xi_p + \delta$, which completes the proof of the lemma.

Next we introduce the new dependent variables

$$\bar{u} = \tfrac{1}{2}\nu u^2 \text{ and } \bar{f} = \tfrac{1}{2}\nu f^2. \tag{17}$$

Then $\bar{u}$ and $\bar{f}$ satisfy in $S^p$ the equations

$$\bar{u}_\xi = \nu u\, \bar{u}_{\eta\eta} + \tfrac{1}{2}\eta \bar{u}_\eta ,$$

and

$$0 = \nu f\, \bar{f}_{\eta\eta} + \tfrac{1}{2}\eta \bar{f}_\eta .$$

If we subtract the equations and write $v = \bar{u} - \bar{f}$, we obtain

$$v_\xi = a\, v_{\eta\eta} + \tfrac{1}{2}\eta v_\eta + cv \quad \text{in } S^p , \tag{18}$$

where $a = a(\xi,\eta) = \nu u(\xi,\eta)$ and $c = c(\eta) = \nu \bar{f}_{\eta\eta}$.

From Lemmas 2 and 4 we can see that for each $\delta > 0$ there exists a constant $\kappa$, which depends on $\delta$, such that

$$a + |a_\eta| + |c| + |c_\eta| \le \kappa < \infty$$

for all $\eta \in \mathbb{R}$ and for all $\xi \ge \xi_p + \delta$. Moreover it follows from Lemma 2 and the asymptotic behaviour of $f(\eta)$ as $|\eta| \to \infty$ that $v$ is uniformly bounded for $\eta \in \mathbb{R}$ and $\xi \ge 0$ and that

$$|\eta|^r\, v(\xi,\eta) \to 0 \quad \text{as } |\eta| \to \infty \tag{19}$$

for any $r \in \mathbb{R}$ and for all $\xi \ge 0$.

LEMMA 5. *Let $u(\xi,\eta)$ be a solution of (12),(13), in which $u_0$ satisfies A1' and A2', and let $f(\eta)$ be the corresponding similarity solution. Let $\bar{u}$ and $\bar{f}$ be defined by (17) and $v = \bar{u} - \bar{f}$. Then for any $\delta > 0$ and $\gamma \in (0,1]$ we have*

$$\sup_{R(\rho,\sigma)} |v_\eta| \le C_2 \{\sup_{R_\delta(\rho,\sigma)} |v|\}^{1-\gamma} \qquad \rho \in \mathbb{R},\ \sigma \ge \xi_p + \delta,$$

*where $C_2$ is a constant which depends only on $\gamma$, $\delta$, $\Delta(p)$ and $\kappa$.*

PROOF. As in the proof of Lemma 4, we obtain for solutions of equation (18)

$$\sup_{R(\rho,\sigma)} |v_\eta| \le C(1 + |\rho|^{\frac{1}{2}}) \sup_{R_\delta(\rho,\sigma)} |v| , \tag{20}$$

where the constant $C$ only depends on $\delta$, $\kappa$ and $\Delta(p)$.

From (19) it follows that for any $\gamma > 0$, there exists a constant $C(\gamma)$ such that

$$(1 + |\rho|)^{\frac{1}{2}} \{ \sup_{R_\delta(\rho,\sigma)} |v|\}^\gamma \le C(\gamma) \quad \text{for all } \rho \in \mathbb{R} \tag{21}$$

and uniformly with respect to $\sigma \geq \xi_p + \delta$. Then combining (20) and (21) yields the desired inequality.

The estimate from Lemma 5 involves rectangles and is therefore not yet the form in which we shall need it. As will become clear in section 5, we shall want an estimate for $|v_\eta|$ on any line $\xi = \xi_0 > \xi_p$ in terms of the supremum of $|v|$ over the line $\xi = \xi_0 - \delta$, where $\delta$ may be any positive number in the interval $(0, \xi_0 - \xi_p)$. Let us denote

$$||\Phi(\xi)|| = \sup_{\mathbb{R}} |\Phi(\xi,\eta)|.$$

LEMMA 6. *Let v be defined as in Lemma 5 and let $\delta$ and $\gamma$ be positive constants such that $\gamma \in (0,1]$. Then there exists a constant L, which depends only on $\delta$, $\gamma$, $\Delta(p)$ and $\kappa$, such that*

$$||v_\eta(\xi)|| \leq L||v(\xi-\delta)||^{1-\gamma}$$

*for any $\xi \geq \xi_p + \delta$.*

PROOF. Let $\xi_0 \geq \xi_p + \delta$ and let $\eta_0 \in \mathbb{R}$ be chosen so that

$$\sup_{\mathbb{R}} |v_\eta(\xi_0,\eta)| = |v_\eta(\xi_0,\eta_0)|.$$

By (19) and Lemma 5, $v_\eta(\xi_0,\eta) \to 0$ as $|\eta| \to \infty$. Hence $\eta_0$ exists. Clearly

$$|v_\eta(\xi_0, \eta_0)| \leq \sup_{R(\eta_0,\xi_0)} |v_\eta|.$$

So by Lemma 5

$$||v_\eta(\xi_0)|| \leq C_2 \{ \sup_{R_\delta(\eta_0,\xi_0)} |v|\}^{1-\gamma}. \tag{22}$$

We complete the proof by using a maximum principle argument. Let $||v(\xi_0 - \delta)|| = M$. Then in view of (19), there exists a number $N > 0$ such that

$$|v(\xi,\eta)| \leq M \quad \text{for} \quad |\eta| \geq N \quad \text{and} \quad \xi \geq \xi_0 - \delta.$$

In the rectangle $(-N,N) \times (\xi_0 - \delta, \xi_0 + 1)$, the coefficients in equation (18) are bounded, and it follows from the maximum principle that

$$|v(\xi,\eta)| \le M\, e^{\kappa(1+\delta)} \quad \text{on} \quad |\eta| \le N,\ \xi_0-\delta \le \xi \le \xi_0+1 \quad . \ (23)$$

Finally, combining (22) and (23) we obtain Lemma 6.

## 5. Proof of Theorem 1.

In section 3 we showed that a solution of Problem II converges, as $x \to \infty$, towards the similarity solution $f(\eta)$, and we obtained an integral estimate for the rate of convergence. In this section we shall use the regularity property (Lemma 6) to derive a pointwise estimate for the rate of convergence, and then we shall return to the original independent variables x and y to obtain Theorem 1.

From Lemma 2 we know that there exists a constant $L_0$, which depends on $u_0$, such that

$$|v| = \tfrac{1}{2}\nu(u+f)|u-f| \le L_0|u-f|.$$

Hence by Theorem 3:

$$\int_{-\infty}^{+\infty} |v(\xi,\eta)|d\eta \le K\, e^{-\xi/2} \quad \text{for all } \xi \ge 0 \ , \qquad (24)$$

where we absorbed the constant $L_0$ in the constant K.

Next consider the following observation.

LEMMA 7. *Let the function $\phi(x)$ be defined and continuously differentiable on $\mathbb{R}$, and let (i) $\phi(x) \ge 0$ for $x \in \mathbb{R}$, (ii) $|\phi'(x)| \le 1$ for $x \in \mathbb{R}$, and (iii) $\int_{-\infty}^{+\infty} \phi(x)dx \le \alpha$. Then*

$$||\phi|| = \sup_{\mathbb{R}} |\phi(x)| \le (\alpha 1)^{\frac{1}{2}}.$$

Since the proof is given in [2], we shall omit it here.

We shall proceed as follows.

First assume without loss of generality that $v \ge 0$ on $\bar{S}$. By Lemma 2, v is uniformly bounded on $\bar{S}$, which implies by Lemma 6 that $|v_\eta|$ is uniformly bounded on $S^{p+\delta}$. So the first two conditions of Lemma 7 are satisfied for $\xi \ge \xi_p + \delta$. The third one is supplied by (24). Therefore, we may conclude that

$$||v(\xi)|| \leq ||v_\eta(\xi)||^{\frac{1}{2}} (K\, e^{-\xi/2})^{\frac{1}{2}} \text{ for } \xi \geq \xi_p + \delta \qquad (25)$$

However, this estimate of $v(\xi,\eta)$ can again be used in Lemma 6 to obtain a new estimate for $|v_\eta(\xi,\eta)|$ for $\xi \geq \xi_p + 2\delta$, which in turn can be used in (25) to give a second estimate for $v(\xi,\eta)$ for $\xi \geq \xi_p + 2\delta$. Now note that this second estimate has a higher exponent and therefore yields a faster decay rate. By repeating this process a sufficient number of times we can arrive at a decay rate which is arbitrary close to $O(e^{-\xi/2})$.

Remembring that $\xi = \log(1 + x)$, we obtain (for the proof, see [2], Theorem 5).

THEOREM 4. *Let $u(x,\psi)$ be the solution of Problem II, in which $u_0$ satisfies A1' and A2'. Then for each $\varepsilon\in(0,1]$, there exists a constant $K(\varepsilon)$ such that*

$$\sup_{\psi\in\mathbb{R}} |u(x,\psi) - f(\psi(x+1)^{-\frac{1}{2}})| \leq K(\varepsilon)(x+1)^{-\frac{1}{2}(1-\varepsilon)} \text{ for all } x \geq 0.$$

*Here the constant $K(\varepsilon)$ depends in addition on $\kappa$, $\Delta(p)$ and $u_0$.*

Now we turn to the physical variables x and y. Let $\bar{u}(x,y) = f(\psi(x,y).(x+1)^{-\frac{1}{2}})$ be the similarity solution which corresponds to Problem II, regarded as a function of x and y, and let (x,y) be an arbitrary point in $\bar{Q}$. Then

$$|u(x,y) - \bar{u}(x,y)| = |u(x,\psi_u) - f(\psi_{\bar{u}}.(x+1)^{-\frac{1}{2}})|, \qquad (26)$$

where

$$\psi_u = \int_0^y u(x,s)ds \text{ and } \psi_{\bar{u}} = \int_0^y \bar{u}(x,s)ds.$$

Adding and subtracting a term $f(\psi_u.(x+1)^{-\frac{1}{2}})$ in the right hand side of (26) yields

$$|u(x,y)-\bar{u}(x,y)| \leq |u(x,\psi_u)-f(\psi_u.(x+1)^{-\frac{1}{2}})|+|f(\psi_u.(x+1)^{-\frac{1}{2}})-f(\psi_{\bar{u}}.(x+1)^{-\frac{1}{2}})|$$

Now using Theorem 4 we obtain

$$|u(x,y)-\bar{u}(x,y)| \leq K(\varepsilon)(x+1)^{-\frac{1}{2}(1-\varepsilon)} + \sup_{\eta\in\mathbb{R}}|f_\eta|.(x+1)^{-\frac{1}{2}}.|\psi_u-\psi_{\bar{u}}|$$

$$\leq K(\varepsilon)(x+1)^{-\frac{1}{2}(1-\varepsilon)} + \sup_{\eta\in\mathbb{R}}|f_\eta|.(x+1)^{-\frac{1}{2}}.\int_0^y |u(x,s)-\bar{u}(x,s)|ds.$$

If we apply Gronwall's Lemma to this inequality, we obtain

$$|u(x,y)-\bar{u}(x,y)| \leq K(\varepsilon)(x+1)^{-\frac{1}{2}(1-\varepsilon)} \cdot e^{K_0 y.(x+1)^{-\frac{1}{2}}},$$

for all $(x,y)\in\bar{Q}$, where $K_0 = \sup_{\eta\in\mathbb{R}} |f_\eta|$.

## REFERENCES


1. CRAVEN, A.H., & L.A. PELETIER, Similarity solutions for degenerate quasilinear parabolic equations, J.Math.Anal.Appl., 38 (1972), 73-81.

2. DUYN, C.J. van, & L.A. PELETIER, Asymptotic behaviour of solutions of a nonlinear diffusion equation, Arch.Rat.Mech.Anal., 65 (1977), 363-377.

3. DUYN, C.J. van, & L.A. PELETIER, A class of similarity solutions of the nonlinear diffusion equation, Nonlinear Analysis, Theory, Methods and Applications, 1 (1977), 223-233.

4. MEYER, R.E., Introduction to Mathematical Fluid Dynamics, New York: Wiley-Interscience 1971.

5. OLEINIK, O.A., The Prandtl system of equations in boundary layer theory, Dokl.Akad.Nauk.SSSR, 150 (1963), 28-32 [English trans. Sov. Math. 4 (1963), 583-586].

6. OLEINIK, O.A., & S.N. KRUZHKOV, Quasilinear second order parabolic equations with many independent variables, Uspekhi Matem. Nauk., 16 (1961), 115-155 [English transl. Russ.Math.Surveys 16 (1961), 105-146].

7. OLEINIK, O.A., A.S. KALASHNIKOV & CHZHOU YUI-LIN, The Cauchy problem and boundary problems for equations of the type of non-stationary filtration, Izv.Akad.Nauk.SSSR Ser. Mat. 22 (1958), 667-704.

8. PELETIER, L.A., Asymptotic behaviour of temperature profiles of a class of non-linear heat conduction problems, Quart.J.Mech.Appl.Math., 23 (1970), 441-447.

9. PELETIER, L.A., On the asymptotic behaviour of velocity profiles in laminar boundary layers, Arch.Rat.Mech.Anal., 45 (1972), 110-119.

10. SCHLICHTING, H., Boundary layer theory, 6th Ed. New York: McGraw-Hill 1968.

11. SERRIN, J., Asymptotic behaviour of velocity profiles in the Prandtl boundary layer theory, Proc.Roy.Soc., A 299 (1967), 491-507.

*Differential Equations and Applications*
*W. Eckhaus and E.M. de Jager (eds.)*
*©North-Holland Publishing Company (1978)*

# ON A CLASS OF PARTIAL FUNCTIONAL DIFFERENTIAL EQUATIONS ARISING IN FEED-BACK CONTROL THEORY.

A. van Harten and J.M. Schumacher
Mathematisch Instituut, Rijksuniversiteit, Utrecht.
Wiskundig Seminarium, Vrije Universiteit, Amsterdam.

Abstract.
In this paper we consider systems of diffusion type controlled by an instantaneous feed-back mechanism based on a finite number of permanent observations of the state-variable. Attention is paid to the existence, uniqueness, regularity and continuous dependence on data of a solution of the IBVP for the non-local, $2^{nd}$ order, parabolic PFDE, governing the evolution in time of this controlled system. Further the question of stability of stationary states of the controlled system is considered.

## 1. Mathematical formulation of the problem.

Let D be a bounded domain $\subset \mathbb{R}^N$ with a $C^\infty$ boundary $\partial D = \overline{D}\setminus D$. $Q_T \subset \mathbb{R}^{N+1}$ will be the cylinder $D\times(0,T)$. The behaviour of the controlled system is described by $u(x,t)$ with $(x,t) \in \overline{Q}_T$, where u has to satisfy:

$$(1.1)\qquad \begin{array}{ll} \dfrac{\partial u}{\partial t} = (L + \Pi)u + f & \text{in } \overline{Q}_T \\ Bu = \phi & \text{BC on } \overline{\Gamma} = \partial D\times[0,T] \\ u(\cdot\,,o) = \psi & \text{IC} \end{array}$$

L will be a linear, 2nd order, uniformly elliptic, PDO with time-independent coefficients $\in C^\infty(\overline{D})$. B is an operator of order $\nu$, $\nu = 0$ or $\nu = 1$, of the following form:

$$(1.2)\qquad \begin{array}{ll} B \equiv 1 & \text{if } \nu = 0 \\ B = \sum_{i=1}^{N} b_i \dfrac{\partial}{\partial x_i} + b_0 & \text{if } \nu = 1 \end{array}$$

with time-independent coefficients $b_i \in C^\infty(\partial D)$, $i = 0,\dots,N$. Further $\sum_{i=1}^{N} b_i n_i > 0$ everywhere on $\partial D$, where n denotes the

outward-directed normal on $\partial D$.
$\Pi$ is the feed-back control operator:

$$(1.3) \qquad \Pi = \sum_{i=1}^{p} c_i P_i$$

The $P_i$'s are called observators and the $c_i$'s are called control functions. Observators as well as control functions are taken time-independent.
Let $C^{\alpha}(\overline{D})$ be the Hölderspace of order $\alpha \geq 0$ with its usual norm: $|\ |_{\alpha}$. By $\mathcal{D}_{\alpha}(\overline{D})$ we denote the dual space of $C^{\alpha}(\overline{D})$, i.e. the space of continuous linear functionals on $C^{\alpha}(\overline{D})$ (distributions of order $\alpha$ on $\overline{D}$).
Now we suppose:

$$(1.4) \qquad P_i \in \mathcal{D}_{\alpha}(\overline{D}) \qquad i = 1,\ldots,p \quad \text{with } \alpha \geq 0$$

The control functions $c_i$ will be elements of $C^{\gamma}(\overline{D})$ for a certain $\gamma \geq 0$.
Note, that $\Pi$ is an operator from $C^{\alpha}(\overline{D}) \to C^{\gamma}(\overline{D})$, with a finite ($\leq p$) dimensional range and of a non-local character, if $\Pi \neq 0$.
Let $C^{\beta,\beta/2}(\overline{Q}_T)$ be the Hölderspace of order $\beta$ in x-direction and of order $\beta/2$ in t-direction as introduced in Ladyzenskaja, Solonnikov, Ural'çeva, '67 with its usual norm $|\ |_{\beta,\beta/2}$.
Analogously we introduce $C^{\hat{\beta},\hat{\beta}/2}(\overline{\Gamma})$ with its usual norm $|\ |_{\hat{\beta},\hat{\beta}/2}$. The inhomogeneous terms $f$, $\phi$, $\psi$ will be elements of the respective spaces $C^{\beta,\beta/2}(\overline{Q}_T)$, $C^{\hat{\beta},\hat{\beta}/2}(\overline{\Gamma})$, $C^{\beta_0}(\overline{D})$ with certain $\beta$, $\hat{\beta}$, $\beta_0 \geq 0$. Of course we shall have to make some assumptions on $\beta$, $\hat{\beta}$, $\beta_0$ and also on $\gamma$ in relation to $\alpha$, but this will be done further on.

## 2. A physical example.

Let $u$ represent the distribution of temperature in D. L describes diffusion of heat, convection of heat, exchange of heat with the surroundings (not via the boundary). f represents the autonomous production or absorption of heat. B describes the heat transport through the boundary (i.e. BC as in case ii, $\frac{\partial u}{\partial n} + b_0 u = \phi$). Now suppose $u \equiv 0$ is considered to be an ideal situation. In order to correct for disturbances from this ideal situation one applies the following mechanism.

Temperature is permanently observed in the points $y_1,\dots,y_p$ and this information is instantaneously fed back to a heating/cooling apparatus, which produces/absorbes heat according to the following rule: the amount of heat supplied (+ or -) to a volume element dx at x during a time interval dt at time t as a response to the $i^{th}$ observation is: $c_i(x)\ u(y_i,t)dxdt$. This mechanism gives rise to a control operator:

$$(2.1) \qquad \Pi = \sum_{i=1}^{p} c_i \delta_{y_i}$$

with observators $\delta_{y_i} \in \mathcal{D}_0(\bar{D})$, $\delta_{y_i} v = v(y_i)$ for $v \in C(\bar{D})$.
A sketch of this situation is given below.
However the formulation of §1 also allows us to take observators, such as:

$$(2.2) \qquad (P_i u)(t) = \int_{V_i} u(\xi,t) m_i(\xi) d\xi \text{ with } \int_{V_i} m_i(\xi) d\xi = 1$$

i.e. $P_i$ observes a weighed average over $V_i \subset \bar{D}$, $P_i \in \mathcal{D}_0(\bar{D})$, if $V_i$ open and $m_i \in L_1(V_i)$, or:

$$(2.3) \qquad (P_i u)(t) = \int_{S_i} \langle R_i(\xi) \text{ grad } u(\xi,t), n_{S_i}(\xi) \rangle \, dS_i$$

which contains the case, that $P_i$ observes the total flow of heat through the N-1 dimensional surface $S_i$, $P_i \in \mathcal{D}_1(\bar{D})$, if $S_i$ consists of a finite number of $C^1$ parts and if the matrix function $R_i \in \{L_1(S_i)\}^{N\times N}$.
Even observators using still higher order derivatives are allowed !

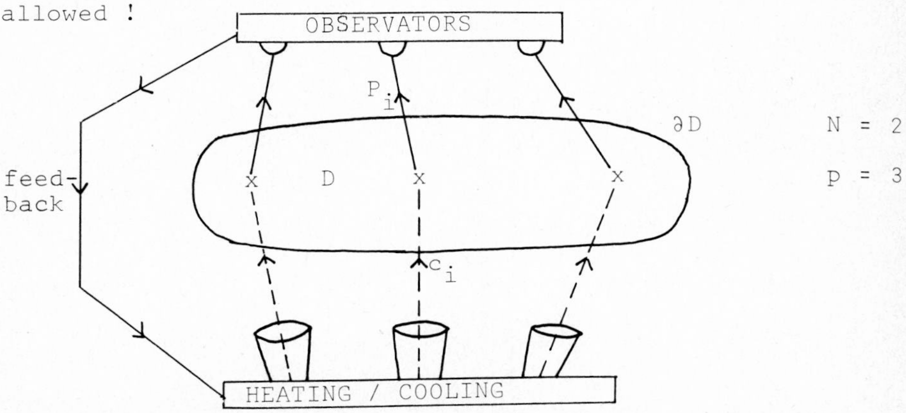

3. Existence, uniqueness, regularity and continuous dependence on data of a solution of the IBVP 1.1 .

Let us first introduce some notation:

(3.1) $s = \max(2, 2\alpha-2, 3\alpha-4-\nu)$

For any number $z \in \mathbb{R}$ we denote by $[z]$ the largest integer $\leq z$. Now we define:

(3.2) $r(z) = \min\left(\left[\frac{z+2-\alpha}{2}\right], \left[\frac{z-\nu}{2}\right]\right)$

(3.3) $q(z) = \min(z, z+2(2-\alpha)+\nu, z+2-\alpha)$

(3.4) $l(z) = \min(z, z+2-\alpha+\nu)$

In order to prove a result as indicated in the title of this subsection we have to make some assumptions. The first one concerns the regularity of the data and the control functions.

Assumption 1.

(3.5) $f \in C^{\beta,\beta/2}(\bar{Q}_T)$, $\phi \in C^{\hat{\beta},\hat{\beta}/2}(\bar{\Gamma})$, $\psi \in C^{\beta_0}(\bar{D})$

$c_i \in C^{\gamma}(\bar{D})$, $i = 1,\ldots,p$

with: $\gamma \geq s-2$; $\gamma \notin \mathbb{Z}$ and for $\alpha > 2$ also $\gamma-\alpha$, $\gamma-2\alpha \notin \mathbb{Z}$

$\beta = \gamma$

$\hat{\beta} = l(\gamma) + 2-\nu$

$\beta_0 = \gamma + 2$

Note: we require "$\gamma \notin \mathbb{Z}$ and ..." in order to avoid some exceptional cases, which however can be dealt with.

The second assumption concerns the compatibility of the data at $t = 0$.

Let us denote:

(3.6) $u^{(k)} = (D_t^k u)|_{t=0}$, $\phi^{(k)} = (D_t^k \phi)|_{t=0}$

$u^{(k)}$ can be found from the recursion:

(3.7) $u^{(0)} = \psi$

$$u^{(k)} = (L+\Pi)u^{(k-1)} + (D_t^{k-1}f)|_{t=0}$$

Note, that $u^{(k)}$ is defined for $0 \leq k \leq r(\beta_0)$ and that $u^{(k)} \in C^{\beta_0-2k}(\overline{D})$ with $\beta_0 - 2k \geq \nu$.

Assumption 2.

(3.8) $\quad Bu^{(k)} = \phi^{(k)}$ on $\partial D$ for $0 \leq k \leq r(\beta_0)$

Besides of these assumptions 1 and 2, which are rather analogous to assumptions made in the uncontrolled case, see Ladyzenskaja, Solonnikov, Ural'çeva, '67, we have to make an additional assumption in the case where the order of the feed-back control operator ($\alpha$) is larger than or equal to the order of the uncontrolled operator (2).

Let $\mathcal{C}_j$ be the solution of:

(3.9) $\quad \frac{\partial \mathcal{C}_j}{\partial t} = L\mathcal{C}_j$

$B\mathcal{C}_j = 0 \qquad$ BC on $\overline{\Gamma}$

$\mathcal{C}_j(\cdot,0) = c_j \qquad$ IC

It is well-known that $\mathcal{C}_j \in L_2(Q_T) \cap C^{\infty}(\overline{D} \times (0,T])$.

For $\alpha < 2$ we can show, that $\mathcal{C}_j \in L_1((0,T) \to C^{\alpha}(\overline{D}))$, see Van Harten, Schumacher, [2].

Now our 3rd assumption is that, if $\alpha \geq 2$, the control functions are chosen, such that:

Assumption 3.

(3.10) $\quad \mathcal{C}_j \in L_1((0,T) \to C^{\alpha}(\overline{D}))$

This 3rd assumption really restricts the choice of the control functions, if $\alpha \geq 2$. Further on (section 5) we shall show, that for $\alpha \geq 2$ some kind of restriction on the choice of the $c_j$'s is indeed necessary in order to avoid pathological counter examples. It be noticed, that 3.10 is certainly satisfied for control functions $c_j \in C^{\infty}(\overline{D})$ with a compact support $\subset D$.

THEOREM 3.I.
*Suppose that assumptions* 1, 2 *and* 3 *are satisfied. Define:*

(3.11) $\tilde{\gamma} = q(\beta_0)$

*There exists a unique solution* u *of the* IBVP 1.1 *in the space* $C^{\tilde{\gamma},\tilde{\gamma}/2}(\bar{Q}_T)$, *which depends continuously on the data in the following sense:*

(3.12) $|u|_{\tilde{\gamma},\tilde{\gamma}/2} \leqslant K\{|f|_{\beta,\beta/2} + |\phi|_{\hat{\beta},\hat{\beta}/2} + |\psi|_{\beta_0}\}$

*with a constant* $K > 0$ *only dependent on* L, Π.

Proof of theorem 3.I.
To start with we have to introduce some more notation:
$\bar{\gamma}_0 = l(\beta_0)$, $\bar{\gamma} = \bar{\gamma}_0 - 2$, $\hat{\bar{\gamma}} = \bar{\gamma}_0 - \nu = \hat{\beta}$, $\mu = \frac{1}{2}(\bar{\gamma}_0 - \alpha)$
$C_0^{2\delta,\delta}(\bar{Q}_T) = \{g \in C^{2\delta,\delta}(\bar{Q}_T) | (D_t^k g)|_{t=0} = 0,\ 0 \leqslant k \leqslant [\delta]\}$
$C_0^{2\delta,\delta}(\bar{\Gamma})$ and $C_0^{\delta}[0,T]$ are defined analogously.
First we introduce $w = u-v$ as a new dependent variable, where $v \in C^{\bar{\gamma}_0,\bar{\gamma}_0/2}(\bar{Q}_T)$ is chosen such, that $(D_t^k v)|_{t=0} = u^{(k)}$ for $0 \leqslant k \leqslant r(\beta_0)$. The existence of such a function v is ensured with $|v|_{\bar{\gamma}_0,\bar{\gamma}_0/2} \leqslant K_1(|f|_{\beta,\beta/2} + |\psi|_{\beta_0})$. The problem for w then becomes:

(3.13)
$$\frac{\partial w}{\partial t} = (L + \Pi)w + f_0$$
$$Bw = \phi_0 \qquad \text{BC on } \bar{\Gamma}$$
$$w(\cdot\, 0) = 0 \qquad \text{IC}$$

with $f_0 \in C_0^{\bar{\gamma},\bar{\gamma}/2}(\bar{Q}_T)$ and $\phi_0 \in C_0^{\hat{\beta},\hat{\beta}/2}(\bar{\Gamma})$.

Secondly we rewrite 3.13 as an equivalent integral equation: $w \in C^{\tilde{\gamma},\tilde{\gamma}/2}(\bar{Q}_T)$ satisfies 3.13 if and only if:

(3.14)
$$w(x,t) = W(x,t) + \sum_{j=1}^{p} \int_0^t \mathcal{C}_j(x,t-\tau)P_j w(\cdot,\tau)d\tau$$

Here W denotes the solution of the uncontrolled problem corresponding to 3.13. It follows from the theory of IBVP for 2nd order parabolic PDE as given in L.S.U. '67, that

$W \in \overset{0}{C}{}^{\bar{\gamma}_0,\bar{\gamma}_0/2}(\bar{Q}_T)$ and $|W|_{\bar{\gamma}_0,\bar{\gamma}_0/2} \leq K_2(|f_0|_{\bar{\gamma},\bar{\gamma}/2} + |\phi_0|_{\hat{\beta},\hat{\beta}/2})$.

Thirdly it appears to be possible to reduce 3.14 to a Volterra integral equation in $\mathbb{R}^p$. Let us introduce:

$\chi(t)$: the p-vector function with components $P_i w(\cdot,t)$

$\eta(t)$: " " " " " $P_i W(\cdot,t)$

$H(t)$: the p×p-matrix function with entries $P_i \mathcal{C}_j(\cdot,t)$

Operating at both sides of 3.14 with $P_i$ and using the fact, that $P_i$ can be pulled to the right side of the integral sign ($P_i \int_0^t = \int_0^t P_i$), since $\mathcal{C}_j \in L_1((0,T) \to C^\alpha(\bar{D}))$ (assumption 3!), we find that $\chi$ has to satisfy:

$$(3.15) \qquad \chi(t) = \eta(t) + \int_0^t H(t-\tau)\chi(\tau)d\tau$$

with $\eta \in \{\overset{0}{C}{}^\mu[0,T]\}^p$ and $H \in \{L_1[0,T]\}^{p\times p}$.

It is rather easy to verify, that 3.15 has a unique solution $\chi \in \{\overset{0}{C}{}^\mu[0,T]\}^p$ with $|\chi|_\mu \leq K_3|\eta|_\mu$.

Finally we note, that $Q(x,t) = \sum_{j=1}^{p} \int_0^t \mathcal{C}_j(x,t-\tau)\chi_j(\tau)d\tau$ with $\chi$ the solution of 3.15 satisfies:

$$(3.16) \qquad \frac{\partial Q}{\partial t} = LQ + \sum_{j=1}^{p} c_j(x)\chi_j(t)$$

$$BQ = 0 \qquad \text{BC on } \bar{\Gamma}$$

$$Q(\cdot,0) = 0 \qquad \text{IC}$$

and as a consequence $Q \in \overset{0}{C}{}^{\tilde{\gamma},\tilde{\gamma}/2}(\bar{Q}_T)$ and $|Q|_{\tilde{\gamma},\tilde{\gamma}/2} \leq K_4|\chi|_\mu$

Now:

$$(3.17) \qquad w = W + Q \in \overset{0}{C}{}^{\tilde{\gamma},\tilde{\gamma}/2}(\bar{Q}_T)$$

is the unique solution of 3.14 and the proof of th. 3.I is easily completed. ∎

Note that for $\alpha < 2$ we have $\tilde{\gamma} = \beta_0 > 2$ i.e. the solution doesn't loose regularity in the x-direction for $t > 0$. However, for $\alpha \geq 2$ it is possible, that $2 < \tilde{\gamma} < \beta_0$, which means that a loss of regularity in the x-direction takes place for $t > 0$. Sometimes this latter situation can be highly undesirable. Fortunately, in the case:

(3.18) $f \equiv 0, \phi \equiv 0$

we can prove a better regularity result for $t > 0$ under somewhat stronger conditions than before.

Assumption 1'.

(3.19) $\psi \in C^{\beta_0}(\overline{D})$; $c_i \in C^{\gamma}(\overline{D})$ $i = 1,\dots,p$
$\beta_0 \geqslant s$; $\beta_0, \beta_0 - \alpha, \beta_0 - 2\alpha \notin \mathbb{Z}$
$\gamma \geqslant \beta_0 - 2$; $\gamma \notin \mathbb{Z}$

Assumption 2'.

(3.20) $B(L + \Pi)^k \psi = 0$ on $\partial D$ for $0 \leqslant k \leqslant r(\beta_0)$

Assumption 3'.

The controlfunctions $c_j$ are such that:

(3.21) $\mathcal{C}_j \in L_1((0,T) \to C^{\alpha}(\overline{D}))$ $1 \leqslant j \leqslant p$

(3.22) $P_i\mathcal{C}_j \in C^{\delta}[0,T]$ $1 \leqslant i,j \leqslant p$

with $\delta \geqslant 0$.

T H E O R E M 3.II.

*Suppose that in the case* 3.18 *assumption* 1', 2' *and* 3' *are satisfied. Define:*

(3.23) $\gamma' = \min(\gamma+2, 2([\frac{q(\beta_0)}{2}] + [\delta] + \max(\frac{q(\beta_0)}{2} - [\frac{q(\beta_0)}{2}], \delta - [\delta])))$

*For each* $t_0 \in (0,T_0)$ *the solution* $u$ *is in* $C^{\gamma',\gamma'/2}(\overline{D} \times [t_0,T])$ *and :*

(3.24) $|u|_{\gamma',\gamma'/2 \text{ on } \overline{D}\times[t_0,T]} \leqslant C(t_0,T)|\psi|_{\beta_0}$

Proof of th. 3.II: see van Harten, Schumacher, [2].

Note, that for $\gamma' \geqslant \beta_0$ no loss of regularity in the x-direction occurs for $t > 0$ !

4. The stationary problem.

Here we consider the elliptic BVP:

(4.1) $(L + \Pi - \lambda)w = g$

$$Bw = 0 \qquad \text{BC on } \partial D$$

When in 1.1 $f$ and $\phi$ are time-independent, then a corresponding stationary solution $u$ of 1.1 will satisfy $(L + \Pi)u + f = 0$, $Bu = \phi$ on $\partial D$. If we introduce $w = u - G\phi$ as a new dependent variable, where the function $G\phi$ satisfies $BG\phi = \phi$ on $\partial D$ and $|G\phi|_{\beta_0} \leq C(\hat{\beta})\ |\phi|_{\hat{\beta}}$, $\beta_0 = \hat{\beta}+\nu$ for all $\hat{\beta}$ for which $\phi \in C^{\hat{\beta}}(\partial D)$, then $w$ will be a solution of 4.1 with $\lambda = 0$, $g = -f-(L+\Pi)G\phi$. The parameter $\lambda \in \mathbb{C}$ introduced in 4.1 has to be considered as a spectral parameter. The topic of this section will be existence, uniqueness, continuous dependence on $g$ of a solution of 4.1. Of course a regularity assumption is made:

<u>Assumption 1"</u>.

(4.2) $\quad g \in C^{\beta}(\overline{D})$, $c_j \in C^{\beta}(\overline{D})$, $1 \leq j \leq p$, $\beta \geq \min(\alpha-2,0)$, $\beta \notin \mathbb{Z}$.

Let $\sigma(L)$ denote the spectrum associated to the uncontrolled problem corresponding to 4.1 ($\Pi \equiv 0$). It is well-known, that $\sigma(L)$ is discrete (Agmon, 1962). As for the problem 4.1 it is relevant to distinguish between the cases <u>a</u>. $\lambda \notin \sigma(L)$, <u>b</u>. $\lambda \in \sigma(L)$.

<u>a</u>. In the case $\lambda \notin \sigma(L)$ the solution of 4.1 with $\Pi \equiv 0$ will be denoted by $(L-\lambda)^{-1}g$. Using Banach's inverse operator theorem, one easily proves that $(L-\lambda)^{-1}$ is a bounded operator from $C^{\beta}(\overline{D})$ onto $C_B^{\beta+2}(\overline{D}) = \{u \in C^{\beta+2}(\overline{D})\,|\, Bu = 0 \text{ on } \partial D\}$.

<u>b</u>. In the case $\lambda \in \sigma(L)$ the operator $L-\lambda$ from $C_B^{\beta+2}(\overline{D})$ into $C^{\beta}(\overline{D})$ is not surjective. Now we have:

(4.3)
$$C_B^{\beta+2}(\overline{D}) = M(\lambda) \oplus \ker(L-\lambda)$$
$$C^{\beta}(\overline{D}) = \text{range }(L-\lambda) \oplus Q_1(\lambda) \oplus \dots \oplus Q_{m(\lambda)}(\lambda)$$

with: $\quad \dim\ker(L-\lambda) \underset{\text{def}}{=} m(\lambda) < \infty$, $\dim Q_i(\lambda) = 1$, $1 \leq i \leq m(\lambda)$, $(L-\lambda)\,M(\lambda) = \text{range }(L-\lambda)$, $(L-\lambda)|_{M(\lambda)}$ injective.

Let $\hat{\phi}_{i,\lambda} \in Q_i(\lambda)$ be $\neq o$, $1 \leq i \leq m(\lambda)$ and let $q_i(\lambda)$

be the projector $C^\beta(\overline{D}) \to Q_i(\lambda)$ according to the decomposition given in 4.3. Define the clf $\hat{q}_i(\lambda)$ by: $q_i(\lambda)\, f = (\hat{q}_i(\lambda) f)\hat{\phi}_{i,\lambda}$.
We define a restricted resolvent in the following way:

$$(4.4) \qquad (L-\lambda)|^{-1} = (L-\lambda)|^{-1}_{M(\lambda)} \; (I - \sum_{i=1}^{m(\lambda)} q_i(\lambda))$$

As before it is easily seen, that $(L-\lambda)|^{-1}$ is a bounded operator from $C^\beta(\overline{D})$ into $C_B^{\beta+2}(\overline{D})$. By $\phi_{i,\lambda}, 1 \leq i \leq m(\lambda)$ we shall indicate a basis of ker $(L-\lambda)$.
Let us introduce some further notation:

a. $\lambda \notin \sigma(L)$. Let $\Omega(\lambda)$ be the p×p-matrix and let $\eta(\lambda) \in \{\mathcal{D}_\beta(\overline{D})\}^p$ be the p-vector of continuous linear functionals on $C^\beta(\overline{D})$, such that:

$$(4.5) \qquad [\Omega(\lambda)]_{i,j} = \delta_{i,j} + P_i(L-\lambda)^{-1} c_j$$

$$(4.6) \qquad [\eta(\lambda) g]_i = P_i(L-\lambda)^{-1} g$$

b. $\lambda \in \sigma(L)$. Let $\hat{\Omega}(\lambda)$ the $\hat{p}\times\hat{p}$-matrix and let $\hat{\eta}(\lambda) \in \{\mathcal{D}_\beta(\overline{D})\}^{\hat{p}}$ be the $\hat{p}$-vector of clf's on $C^\beta(\overline{D})$, such that:

$$(4.7) \qquad \hat{p} = p + m(\lambda)$$

$$(4.8) \qquad [\hat{\Omega}(\lambda)]_{i,j} = \begin{cases} \delta_{i,j} + P_i(L-\lambda)|^{-1} c_j & 1 \leq i,j \leq p \\ P_i\, \phi_{j-p,\lambda} & 1 \leq i \leq p, j > p \\ \hat{q}_{i-p}(\lambda)\, c_j & i > p, 1 \leq j \leq p \\ 0 & i > p, j > p \end{cases}$$

$$(4.9) \qquad [\hat{\eta}(\lambda)]_i = \begin{cases} P_i(L-\lambda)|^{-1} g & 1 \leq i \leq p \\ \hat{q}_{i-p}(\lambda)\, g & i > p \end{cases}$$

Finally we shall denote by $c_\cdot$ the p-vector with components $c_i$ and we shall denote by $\binom{c_\cdot}{0}$, $\binom{0}{\phi_{\cdot,\lambda}}$ $\hat{p}$-vectors with components as indicated.
The inner product in $\mathbb{R}^p$ as well as in $\mathbb{R}^{\hat{p}}$ is denoted by <,>.

T H E O R E M 4.I.
*Suppose, that assumption 1" is valid.*

a. $\lambda \notin \sigma(L)$.

*The BVP 4.1 is uniquely solvable in* $C^{\beta+2}(\bar{D}) \Leftrightarrow \Omega(\lambda)$ *is invertible. If* $\Omega(\lambda)$ *is invertible, then:*

$$(4.10) \qquad u = (L-\lambda)^{-1}(g - \langle c_{\cdot}\,, \Omega(\lambda)^{-1}\eta(\lambda)g\rangle)$$

*and* u *satisfies an estimate:*

$$(4.11) \qquad |u|_{\beta+2} \leqslant K(\lambda)|g|_{\beta}$$

b. $\lambda \in \sigma(L)$.

*The BVP 4.1 is uniquely solvable in* $C^{\beta+2}(\bar{D}) \Leftrightarrow \hat{\Omega}(\lambda)$ *is invertible. If* $\hat{\Omega}(\lambda)$ *is invertible, then:*

$$(4.12) \qquad u = (L-\lambda)|^{-1}(g - \langle \binom{c_{\cdot}}{0}, \hat{\Omega}(\lambda)^{-1}\,\hat{\eta}(\lambda)g\rangle) - \langle \binom{0}{\phi_{\cdot,\lambda}}, \hat{\Omega}(\lambda)^{-1}\,\hat{\eta}(\lambda)g\rangle$$

*and* u *satisfies an estimate as in 4.11.*

c. $\lambda \notin \sigma(L)$ *and* $\Omega(\lambda)$ *singular or* $\lambda \in \sigma(L)$ *and* $\hat{\Omega}(\lambda)$ *singular* $\Leftrightarrow \lambda \in \sigma(L + \Pi) \Leftrightarrow$ *the homogeneous* ($g \equiv 0$) *form of the problem* 4.1 *has non-trivial solutions in* $C^{\beta+2}(\bar{D})$.

Proof of th. 4.I.

a. u is a solution of 4.1 with $\lambda \notin \sigma(L) \Leftrightarrow$

$$(4.13) \qquad \xi_j = P_j u$$

$$(4.14) \qquad u = (L-\lambda)^{-1}\,(g - \langle c_{\cdot}\,,\xi\rangle)$$

Operating at both sides of 4.14 with $P_i$, we find:

$$(4.15) \qquad \Omega(\lambda)\xi = \eta(\lambda)g$$

and conversely, if $\xi$ satisfies 4.15, then u as defined in 4.14 satisfies 4.1.

Now a. and half of c. are easily verified.

b. u is a solution of 4.1 with $\lambda \in \sigma(L) \Leftrightarrow$

$$(4.16) \qquad \hat{\xi}_j = P_j u \qquad j = 1,\ldots,p$$

(4.17) $g - \sum_{j=1}^{p} \hat{\xi}_j c_j \in \text{range } (L-\lambda)$, i.e.

$$\hat{q}_{i-p}(\lambda) g = \sum_{j=1}^{p} \hat{\xi}_j \hat{q}_{i-p}(\lambda) c_j \qquad i = p+1,\dots,\hat{p}$$

(4.18) $$u = (L-\lambda)|^{-1}(g - \sum_{j=1}^{p} \hat{\xi}_j c_j) - \sum_{j=p+1}^{\hat{p}} \hat{\xi}_j \phi_{j-p,\lambda}$$

Operating at both sides of 4.18 with $P_i$ and combining with 4.17, we find:

(4.19) $$\hat{\Omega}(\lambda)\hat{\xi} = \hat{\eta}(\lambda) g$$

and conversely, if $\hat{\xi}$ satisfies 4.19, then u as defined in 4.18 satisfies 4.1.
Now b and the other half of c are easily verified and this completes the proof. ■

Note, that theorem 4.I implies, that for the problem 4.1 Fredholm's alternative holds true.
Further it be noticed, that the regularity condition necessary for theorem 4.I is weaker, than the one used for theorem 3.I, if $\alpha \geq 2$.

## 5. A further characterization of $\sigma(L + \Pi)$.

From now on we shall suppose that the functions $c_i$ satisfy:
Assumption 1"-c:
$\exists\gamma$ such that $\gamma > 0$, $\gamma \geq \alpha-2$, $\gamma \notin \mathbb{N}$ and $c_j \in C^{\gamma}(\bar{D})$, $i \leq j \leq p$.

We shall show, that there exists a nice relation between $\sigma(L + \Pi)$ and the meromorphic function:

(5.1) $$\omega(\lambda) = \det \Omega(\lambda)$$

THEOREM 5.I.

(i) $\omega \equiv 0$ on $\mathbb{C} \Leftrightarrow \sigma(L + \Pi) = \mathbb{C}$

(ii) $\exists\lambda_0 \in \mathbb{C}$ $\omega(\lambda_0) \neq 0 \Leftrightarrow \sigma(L + \Pi)$ *consists of a denumerable set of isolated eigenvalues with finite multiplicities without accumulation points. The following characterization is valid:*

(5.2) $\tilde{\lambda} \notin \sigma(L)$ *and* $\omega(\lambda)$ *has a zero of order* $m > 0$ *at* $\tilde{\lambda}$
$\Leftrightarrow \tilde{\lambda} \in \sigma(L + \Pi)$ *and alg. mult* $(\tilde{\lambda}; L + \Pi) = m > 0$.

(5.3) $\tilde{\lambda} \in \sigma(L)$, *alg. mult* $(\tilde{\lambda};L) = m_0 > 0$ *and* $\omega(\lambda)$ *has Laurent-index* $m > -m_0$ *at* $\tilde{\lambda}$ (*i.e.* $\omega(\lambda) = a_m(\lambda-\tilde{\lambda})^m + O(\lambda-\tilde{\lambda})^{m+1}$ *for* $\lambda \to \tilde{\lambda}$ *with* $a_m \neq 0$) $\Leftrightarrow \lambda \in \sigma(L + \Pi)$ *and alg. mult* $(\tilde{\lambda};\, L + \Pi) = m_0 + m$.

Proof of th. 5.I.

(i) is a consequence of th. 4.I-c.

(ii) $(L + \Pi - \lambda_0)^{-1}$ is compact from $C^\gamma(\bar{D}) \to C^\gamma(D)$. This implies the first part of (ii). The characterizations 5.2-3 follow from the interpretation of $\omega(\lambda)$ as a so-called Weinstein-Aronszajn determinant, see Kato, '66, ch. IV, § 6.2. ■

If $\sigma(L + \Pi) = \mathbb{C}$ it is impossible, that 1.1 possesses a solution, which depends continuously on the data as in theorem 3.I, II. Namely take any $\lambda \in \mathbb{C}$ and let $\psi_\lambda$ be a non-trivial solution of problem 4.1 with $g \equiv 0$ (see theorem 4.I-c). Rewriting equation 4.1 as $(L - \lambda_0)\psi_\lambda = -\Pi\psi_\lambda + (\lambda-\lambda_0)\psi_\lambda$ with $\lambda_0 \notin \psi_\lambda$ and using the A.D.N.*) a-priori estimates repetedly we find, that for each $\varepsilon \in (0,\gamma+2]$ there are constants $C > 0$, $\delta > 0$ only dependent of $\varepsilon$, $\gamma$, L, $\Pi$, such that:

$$|\psi_\lambda|_{\gamma+2} \leqslant C(1 + |\lambda|)^\delta |\psi_\lambda|_\varepsilon \tag{5.4}$$

Now the function $u_\lambda(x,t) = \psi_\lambda(x)\exp(\lambda t)$ satisfies 1.1 with $f \equiv 0$, $\phi \equiv 0$, $\psi = \psi_\lambda$ and:

$$|u_\lambda(\cdot\,,t)|_\varepsilon \geqslant C^{-1}(1 + |\lambda|)^{-\delta} \exp(\mathrm{Re}\,\lambda t)\, |\psi_\lambda|_{\beta+2} \tag{5.5}$$

Since Re $\lambda$ can be chosen arbitrarily large 5.5 clearly contradicts continuous dependence on the data in the following sense: for each $T > 0$ and $\varepsilon > 0$ there exists a sequence $u_n$, $n \in \mathbb{N}$ in $C^{\gamma+2,(\gamma+2)/2}(\bar{Q}_T)$ with $Bu_n = 0$ on $\bar{\Gamma}$, $u_n(\cdot\,,0) \to 0$ for $n \to \infty$ in $C^{\gamma+2}(\bar{D})$, but nót $u_n \to 0$ for $n \to \infty$ in $C^{\varepsilon,\varepsilon/2}(\bar{Q}_T)$.

A natural question is, whether this pathological situation indeed can arise. The answer is yes and this will be demonstrated by the following example.

Example of a case where $\sigma(L + \Pi) = \mathbb{C}$.

Choose $L = \Delta$, $B = \frac{\partial}{\partial n}$, $\Pi = cP$ i.e. $p = 1$,

*): see Agmon, Douglis, Nirenberg, '59

(5.6) $P \in \mathcal{D}_4(\bar{D})$: $Pu = \int_D \Delta\Delta u \, dx$

(5.7) $c \in C^\infty(\bar{D})$, such that $\int_D \Delta c \, dx = -1$

Now $\omega(\lambda) = 1 + P\tilde{c}$ with $\tilde{c}$ the solution of

(5.8) $(\Delta - \lambda)\tilde{c} = c$

$\frac{\partial \tilde{c}}{\partial n} = 0$ BC on $\partial D$

But:

(5.9) $P\tilde{c} = \int_D \Delta c \, dx + \lambda \int_D \Delta \tilde{c} \, dx = -1 + \lambda \int_{\partial D} \frac{\partial \tilde{c}}{\partial n} dS = -1$

$\Rightarrow \omega(\lambda) \equiv 0$

and theorem 5.I-i implies: $\sigma(L + \Pi) = \mathbb{C}$.

However:

T H E O R E M 5.II.

*If* $\alpha < 2$ *and assumption* 1"-c *is satisfied ór if* $\alpha \geq 2$ *and the control functions satisfy:*

*Assumption* 1-3-c.

$\exists\gamma$ *such that* $\gamma > 0$, $\gamma \geq \max(2\alpha-4, 3\alpha-6-\nu)$; $\gamma, \gamma-\alpha, \gamma-2\alpha \notin \mathbb{N}$

*and:* $c_j \in C^\gamma(\bar{D})$, $1 \leq j \leq p$

$\mathcal{C}_j \in L_1((0,T) \to C^\alpha(\bar{D}))$

*then* $\sigma(L + \Pi)$ *is a denumerable set of isolated points without accumulation points.*

Proof of th. 5.II.

The proof is a straightforward combination of theorem 3.I, the conclusion of 5.5 and theorem 5.I. ■

Note, that we can consider assumption 3 of th. 3.I as a condition to avoid pathological cases, such as when $\sigma(L + \Pi) = \mathbb{C}$.

## 6. Stability of stationary solutions, applications of semi-group theory.

In order to investigate the stability of a stationary solution of 1.1 one has to consider the problem:

$$(6.1)\qquad \frac{\partial v}{\partial t} = (L + \Pi)v$$

$$Bv = 0 \qquad \text{BC on } \overline{\Gamma}$$

$$v(\cdot\,,0) = \psi \qquad \text{IC}$$

Let V be a linear space of functions on $\overline{D}$, such that $\psi \in V$ implies, that 6.1 has a unique solution: $v(\cdot\,,t) \underset{\text{def}}{=} T(t)\psi$. Suppose, that there is a normed linear space $\widetilde{V}$ with norm $\|\ \|$, such that $\psi \in V \Rightarrow \exists_{t_0} \geq 0\ \forall_{t \geq t_0}\ T(t)\psi \in \widetilde{V}$.

Then by definition stationary solutions of 6.1 are asymptotically stable for disturbances from V in the sense of $\|\ \|$, abbreviated AS(V,$\|\ \|$), if:

$$(6.2)\qquad \lim_{t \to \infty} \|T(t)\psi\| = 0 \text{ for all } \psi \in V$$

We shall show, that under some conditions $\sigma(L + \Pi) \subset \{\lambda \mid \text{Re}\,\lambda \leq -\delta < 0\}$ implies AS(V,$\|\ \|$) with a rather large space V and a rather strong norm $\|\ \|$, as usual.

6.1. The case $\alpha < 2$.

Again we assume, that assumption 1"-c is valid. Choose $\varepsilon \notin \mathbb{N}$, $\varepsilon \in (0,\gamma)$, $\varepsilon$ arbitrarily small. Let $C^{2+\varepsilon}_{B,r}(\overline{D})$ be $\{\psi \in C^{2+\varepsilon}(\overline{D}) \mid B(L + \Pi)^k\psi = 0 \text{ on } \partial D,\ 0 \leq k \leq r = [\tfrac{1}{2}(2+\varepsilon-\nu)]\}$
$C^{2+\varepsilon}_{B,r}(\overline{D})$ is a Banach space with respect to $|\ |_{2+\varepsilon}$. Because of theorem 3.I $\{T(t) \mid t \geq 0\}$ defines a strongly continuous semigroup of operators on $C^{2+\varepsilon}_{B,r}(\overline{D})$, i.e.

$$(6.1.1)\qquad T(0) = I;\ T(t_1 + t_2) = T(t_1)T(t_2),\ t_1 \geq 0,\ t_2 \geq 0$$

$$\lim_{\substack{\tau \to t \\ \tau \geq 0}} T(\tau)\psi = T(t)\psi \text{ for all } \psi \in C^{2+\varepsilon}_{B,r}(\overline{D})$$

$$|T(t)|_{2+\varepsilon} < \infty \text{ for each } t \geq 0.$$

By definition the infinitesimal generator of this semi-group is the unbounded operator $L + \Pi$ with domain:
$\text{dom}\,(L + \Pi) = \{\psi \in C^{2+\varepsilon}_{B,r}(\overline{D}) \mid (L + \Pi)\psi \in C^{2+\varepsilon}_{B,r}(\overline{D})\}$.
An application of the standard theory of semi-groups at once gives the result:

THEOREM 6.I.

(i) $\sigma(T(t)) = \exp(t\sigma(L + \Pi))$ *possibly* $\cup\{0\}$

(ii) $\hat{\sigma} = \sup_{\lambda \in \sigma(L+\Pi)} \operatorname{Re} \lambda < \infty$

(iii) $\forall_{\hat{\varepsilon} > 0}\ \exists_{K(\hat{\varepsilon}) \geq 1}$ *such that* $\forall_{t \geq 0}$:

(6.1.2) $|T(t)|_{2+\varepsilon} \leq K(\hat{\varepsilon}) \exp((\hat{\sigma} + \hat{\varepsilon})t)$

(iv) $\hat{\sigma} < 0 \Rightarrow$ *stationary solutions of* 1.1 *have the property* $AS(C^{2+\varepsilon}_{B,r}(\bar{D}), |\ |_{2+\varepsilon})$.

Proof of th. 6.I.

For (i) we refer to Hille, Phillips, '57, th. 16.7.2. The contents of (ii) follow directly from (i) and the boundedness of $T(t_0)$ for some $t_0 > 0$. For (iii) we refer, to Hale, '71, lemma 22.2 and (iv) is a direct consequence of (iii). ■

6.2. The case $\alpha \geq 2$.

Now we have to be somewhat more careful in order to ensure, that T(t) defines a semi-group on a suitable space, because of the loss of regularity in x-direction, which can take place for $t > 0$, see section 3. Instead of assumption 1-3-c we take a stronger condition on the controlfunctions $c_j$, which enables us to exploit theorem 3.II.

Assumption (1-3-c)'.

$\exists\gamma$ such that $\gamma > s = \max(2\alpha-2, 3\alpha-4-\nu)$, $\gamma \notin \mathbb{N}$ and:

(6.2.1) $c_j \in C^{\gamma}(\bar{D})$ $\quad 1 \leq j \leq p$

$\mathcal{C}_j \in L_1((0,T) \to C^{\alpha}(\bar{D}))$ $\quad$ "

$P_i \mathcal{C}_j \in C^{\delta}[0,T]$ $\quad 1 \leq i,j \leq p$

with $\delta = 1 + \frac{1}{2}\gamma - [\frac{s}{2}]$

Now choose $\varepsilon \in [0,s-\gamma)$, such that $s+\varepsilon$, $s+\varepsilon-\alpha$, $s+\varepsilon-2\alpha \notin \mathbb{Z}$ and define:

(6.2.2) $C^{s+\varepsilon}_{B,r}(\bar{D}) = \{\psi \in C^{s+\varepsilon}(\bar{D}) \mid B(L + \Pi)^k\psi = 0 \text{ on } \partial D,\ 0 \leq k \leq r(s+\varepsilon)\}$

with r as in 3.2.

$C^{s+\varepsilon}_{B,r}(\bar{D})$ is a Banach space with respect to $|\ |_{s+\varepsilon}$. Under the assumption (1-3-c)' $\{T(t)|t \geq 0\}$ now defines a semi-group of operators on $C^{s+\varepsilon}_{B,r}(\bar{D})$, because of theorem 3.II.
For $t > 0$ this semi-group is strongly continuous, but for $t \downarrow 0$ the behaviour is rather bad! This latter fact makes it difficult to apply the standard semi-group theory. However, the assumption (1-3-c)' is such that for $t > 0$, we know a lot more than strong continuity of the semi-group, namely: the range of $T(t)$ is contained in the domain of the generator $L + \Pi$: $\mathrm{dom}\,(L + \Pi) = \{\psi \in C^{s+\varepsilon}_{B,r}(\bar{D}) | (L + \Pi)\,\psi \in C^{s+\varepsilon}_{B,r}(\bar{D})\}$ and $T(t)$ is a compact operator on $C^{s+\varepsilon}_{B,r}(\bar{D})$.
These properties also follow from theorem 3.II, for that theorem yields $v \in C^{\gamma+2,(\gamma+2)/2}(\bar{D} \times [t_0,T])$ for $t_0 \in (0,T_0)$.
This information is sufficient to deduce the following result:

<u>T H E O R E M 6.II.</u>

(i) $\sigma(T(t)) = \exp\,(t\sigma(L + \Pi))$ *possibly* $\cup \{0\}$

(ii) $\hat{\sigma} = \sup_{\lambda \in \sigma(L + \Pi)} \mathrm{Re}\,\lambda < \infty$

(iii) $\forall \hat{\varepsilon} > 0\ \forall t_0 > 0\ \exists K(\hat{\varepsilon},t_0)$ *such that* $\forall t \geq t_0$

(6.2.3) $|T(t)|_{\gamma+2} \leq K(\hat{\varepsilon},t_0)\exp\,((\hat{\sigma} + \hat{\varepsilon})t)$

(iv) $\hat{\sigma} < 0 \Rightarrow$ *stationary solutions of* 1.1 *have the property* AS $(C^{s+\varepsilon}_{B,r}(\bar{D}), |\ |_{\gamma+2})$.

<u>Proof of th. 6.II.</u>

(i) $\sigma(T(t)) \supset \exp\,(t\sigma(L + \Pi))$ is a trivial consequence of th. 4.I-c.
Suppose $\mu \in \sigma(T(t))$, $\mu \neq 0$, $t > 0$. Let $e$ be an eigenfunction of $T(t)$ for the eigenvalue $\mu$ ($T(t)$ is compact!). Since $T(t)e = \mu e$, we find $e \in \mathrm{dom}(L + \Pi)$ and we have:

$$(6.2.4) \quad T(t)(L + \Pi)e = (L + \Pi)T(t)e = \mu(L + \Pi)e$$

for $L + \Pi$ and $T(t)$ commute on dom $(L + \Pi)$, see Hille, Phillips, '57, th. 10.3.3. Let $N$ be $\ker(T(t) - \mu)$, then $\dim(N) < \infty$ and because of 6.2.4:

(6.2.5) $(L + \Pi)\, N \subset N$

So there exists an element $\tilde{e} \in N$, such that:

(6.2.6) $(L + \Pi)\tilde{e} = \tilde{\mu}\, \tilde{e}$.

But 6.2.6 implies $\tilde{\mu} \in \sigma(L + \Pi)$ and $T(t)\tilde{e} = \tilde{e}\, \exp(\tilde{\mu}\, t)$, so $\mu = \exp\, (\tilde{\mu}\, t) \in \exp(t\, \sigma(L + \Pi))$.
(ii), (iii) and (iv) are proven analogous to the corresponding statements in th. 6.I. ■

7. <u>On the controllability of the location of $\sigma(L + \Pi)$</u>.
Suppose, that for the uncontrolled problem, there exists a point $\lambda \in \sigma(L)$ with $\mathrm{Re}\, \lambda > 0$, i.e. stationary solutions of the uncontrolled system are instable in any reasonable sense. The question now is, whether it is possible to determine the control functions $c_i$ and the observators $P_i$ with $i = 1,\ldots,p$ in such a way, that stationary solutions of the controlled system become asymptotically stable in some sense. We have the following result:

<u>T H E O R E M 7.I.</u>
*There exist* $P_i \in \mathcal{D}_0(\bar{D})$ *and* $c_i \in C^{\infty}(\bar{D})$ *with* $i = 1,\ldots,p$, *such that for a suitable* $\delta > 0$

(7.1) $\sigma(L + \Pi) \subset \{\lambda \,|\, \mathrm{Re}\, \lambda \leqslant -\delta < 0\}$

*if*

(7.2) $\max_{\substack{\lambda \in \sigma(L) \\ \mathrm{Re}\, \lambda \geqslant 0}} [\text{alg. mult } (\lambda; L)] \leqslant p$

For the proof of th. 7.I we refer to van Harten, Schumacher, [2].
Because of th. 6.I we have, that 7.2 implies the existence of $P_i \in \mathcal{D}_0(\bar{D})$ and $c_i \in C^{\infty}(\bar{D})$ with $i = 1,\ldots,p$, such that stationary solutions of the controlled system 1.1 have the property AS $(C^{2+\varepsilon}_{B,r}(\bar{D}),\, |\;|_{2+\varepsilon})$ for each $\varepsilon > 0$, $\varepsilon \notin \mathbb{N}$ !

## REFERENCES

Agmon, S., 1962.
On the eigenfunctions and on the eigenvalues of general elliptic boundary value problems, Comm. Pure & Appl. Math., vol. 15, pg. 119-147.

Agmon, S., Douglis, A., Nirenberg, L., 1959.
Estimates near the boundary for solutions of elliptic partial differential equations satisfying general boundary conditions, I, Comm. Pure & Appl. Math., vol. 12, pg. 623-727.

Hale, J., 1971.
Functional differential equations, Springer, Berlin.

Harten, A. van, Schumacher, J.M., [1].
Some boundary value problems for a class of 2nd order elliptic partial functional differential equations arising in feedback control theory, preprint, report 69, V.U. Amsterdam, 1977.

Harten, A. van, Schumacher, J.M., [2].
preprint, to appear: 1978.

Hille, E., Phillips, R.S., 1957.
Functional analysis and semi-groups, Amer. Math. Soc. Colloq. Publ., vol. 31.

Kato, T., 1966.
Perturbation theory for linear operators, Springer, Berlin.

Ladyzenskaja, O.A., Solonnikov, V.A., Ural'çeva, N.N., 19 .
Linear and quasi-linear equations of parabolic type, Transl. Math. Mon., Am. Math. Soc., vol. 23.

*Differential Equations and Applications*
*W. Eckhaus and E.M. de Jager (eds.)*


# TRANSFORMATION OPERATORS AND WAVE PROPAGATION IN A SPHERICALLY STRATIFIED MEDIUM*

by

David Colton
Department of Mathematics
University of Strathclyde
Glasgow, Scotland.

## I INTRODUCTION

The problem of obtaining constructive methods for solving boundary value problems arising in acoustic and electromagnetic scattering theory has played a dominant role in mathematical physics for almost a century. In recent years a considerable amount of attention has been devoted to the problem of deriving constructive methods for solving wave propagation problems in an inhomogeneous medium. Due to the inherent complexities of such problems, particularly where asymptotic methods are no longer available, much of this work has been concentrated on the special cases of spherical or horizontal stratification (c.f. [3], [15], and the reference cited therein). However even in the case of a stratified medium the methods currently available for approximating solutions to scattering problems at intermediate frequencies are not very satisfactory in the sense that they often require the numerical solution of a coupled system of two and three dimensional integral equations. In the case of a spherically stratified medium, the author, in collaboration with Wolfgang Wendland, Rainer Kress, and George Hsiao, has recently overcome some of these objections through the use of transformation operators for elliptic equations, which are a generalization of the transformation (or translation) operators for ordinary differential equations as developed (among others) by Levitan ([14]), Agranovich and Marchenko ([1]), and Braaksma ([2]). In this paper we shall outline the construction of these transformation operators and show how they can be applied to derive constructive methods for solving boundary value problems arising in the scattering of acoustic waves in a spherically stratified medium. Other applications of transformation operators to problems in wave propagation are also possible, for example, to the problem of radiowave propagation around the earth under the assumption of a spherically stratified atmosphere, but we shall not report on this work at the present time.

## II TRANSFORMATION OPERATORS FOR A CLASS OF ELLIPTIC EQUATIONS

The partial differential equation which arises in the theory of the propagation of acoustic waves in a quasi-homogeneous spherically stratified medium is

$$\Delta_3 u + k^2(1+B(r))u = 0 \tag{2.1}$$

where $k$ is the wave number and $B(r) = \left(\frac{c_o}{c(r)}\right)^2 - 1$, where $c(r)$ is the speed of sound and $c_o = \lim_{r\to\infty} c(r)$. It is assumed that $B(r)$ is continuously differentiable. We shall now introduce two transformation operators which map solutions of the reduced wave equation in a homogeneous medium

* This research was supported in part by AFOSR Grant 76-2879 and NSF Grant MCS 77-02056.

$$\Delta_3 h + k^2 h = 0 \tag{2.2}$$

onto solutions of (2.1). The first of these is valid for interior domains and is related to Gilbert's "method of ascent" ([8]). Its construction is due to Colton, Hsiao, and Kress ([5]) and is of the form

$$u(r,\theta,\phi) = h(r,\theta,\phi) + \int_0^r G(r,s;k)\, h\, (s,\theta,\phi)ds \tag{2.3}$$

where $G(r,s;k)$ is a known kernel whose construction we shall present shortly. Our second transformation operator is valid for exterior domains, is due to Colton, Kress, and Wendland ([6], [7]), and is of the form

$$u(r,\theta,\phi) = h(r,\theta,\phi) + \int_r^\infty E(r,s;k)\, h\, (s,\theta,\phi)ds \tag{2.4}$$

where $E(r,s;k)$ is a known kernel whose existence is assured for $r>b$, $o<k<2\gamma b$, provided

$$B(r) = O(e^{-\gamma r^2}) \quad . \tag{2.5}$$

We shall outline the construction of $E(r,s;k)$ shortly, but first note that the restriction on $k$ and $r$ arises from the fact that the operator defined by (2.4) preserves the far field data, and if $B(r)$ decays more slowly than (2.5)(e.g. exponentially) the far field data of solutions to (2.1) satisfying the Sommerfeld radiation condition are no longer entire as they are for (2.2). Hence one would not expect a relation such as (2.4) to hold.

We now turn to the construction of $G(r,s;k)$ and $E(r,s;k)$. There is a remarkable symmetry in the initial value problem satisfied by these two functions which is illustrated in the figure below. To explain this figure, let the operator $L$ be defined by

$$L = r^2 \frac{\partial^2}{\partial r^2} - s^2 \frac{\partial^2}{\partial s^2} + 2r \frac{\partial}{\partial r} - 2s \frac{\partial}{\partial s} + k^2(r^2 - s^2 + r^2 B(r)) \quad . \tag{2.6}$$

Then $G(r,s;k)$ is a solution of $L(G) = 0$ in the wedge $o<s<r$, $o<r<\infty$, assumes prescribed data on the characteristic $r = s$, and is required to be bounded on the singular line $s = 0$, whereas $E(r,s;k)$ is a solution of $L(E) = 0$ in the wedge $r<s<\infty$, $o<r<\infty$, assumes complimentary prescribed data on the characteristic $r = s$, and is required to decay to zero (exponentially) as $s\to\infty$.

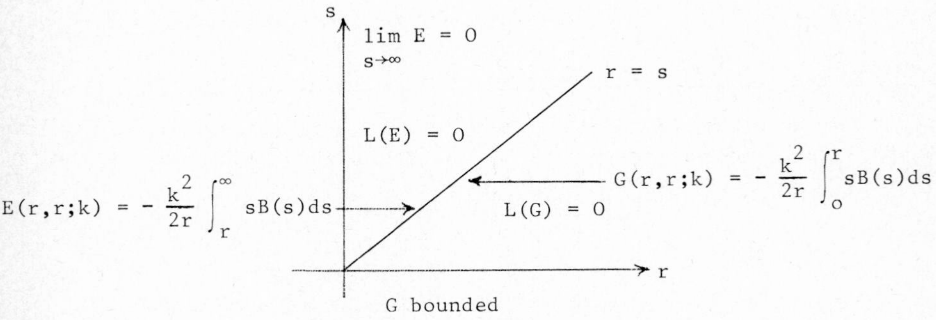

For both of these problems the change of variables

$$\xi = \sqrt{rs}$$

$$\eta \quad \sqrt{\frac{r}{s}}$$

$$M(\xi,\eta;k) = \xi G(\xi\eta,\frac{\xi}{\eta};k) \tag{2.7}$$

$$N(\xi,\eta;k) = \xi E(\xi\eta,\frac{\xi}{\eta};k)$$

reduces the singular hyperbolic equations $L(G) = 0$ and $L(E) = 0$ to ones of normal type, and the initial value problems for M and N can be reformulated as Volterra integral equations over the domains $(0,\xi) \times (1,\eta)$ and $(\xi,\infty) \times (\eta,1)$ respectively ([5], [6]). These integral equations can be solved by successive approximations, and if $B(r)$ is small (with respect to the maximum norm) the first few iterates provide an accurate approximation to the desired kernel.

## III THE SCATTERING OF ACOUSTIC WAVES IN A HOMOGENEOUS MEDIUM

We now wish to apply the transformation operators constructed in Section II to study wave propagation problems in a spherically stratified medium for low and intermediate frequencies and "small" inhomogeneities. However in order to proceed to this task, it is first necessary to briefly examine the methods that have been developed to study scattering problems in a homogeneous medium for low and intermediate values of the frequency. In this case various asymptotic methods are no longer applicable, and the best approach seems to be that of integral equations. However, as is well known, problems arise due to the fact that if the free space Green's function is used as the kernel of the integral operator, the resulting integral equation is no longer invertible at certain "critical values" of the wave number ([13]). In recent years a variety of approaches have been developed to overcome this problem, and for our purposes the methods of Ursell ([16]; see also [10]) and Leis ([13]; see also [11]) are most significant. To describe these methods, let D be a bounded, simply connected domain with smooth boundary $\partial D$ and suppose we want to solve the exterior Neumann problem

$$\Delta_3 h + k^2 h = 0 \text{ in } R^3\backslash D \tag{3.1a}$$

$$\frac{\partial h}{\partial \nu} = f \text{ on } \partial D \tag{3.1b}$$

$$\lim_{r\to\infty} r(\frac{\partial h}{\partial r} - ikh) = 0 \tag{3.1c}$$

where $\nu$ is the (inward pointing) unit normal to $\partial D$ and $f(\underset{\sim}{x})$ is a continuous function prescribed on $\partial D$. Then Ursell's approach is to look for a solution of (3.1a)-(3.1c) in the form

$$h(\underset{\sim}{x}) = \int_{\partial D} \mu(\underset{\sim}{\xi})G(\underset{\sim}{\xi};\ \underset{\sim}{x})\, d\omega_{\xi} \tag{3.2}$$

where $\mu(\xi)$ is a density to be determined and $G(\xi;x)$ is a fundamental solution of the reduced wave equation satisfying the Sommerfeld radiation condition (3.1c) such that on $|x| = b$ (where $\{x: |x| < b\} \subset D$) we have

$$\left(\frac{\partial}{\partial r} + \alpha\right)G = 0; \quad \operatorname{Im}\alpha > 0 . \tag{3.3}$$

$G(\xi;x)$ can be constructed by separation of variables, and if we let $x$ tend to $\partial D$ and apply the boundary condition (3.1b) we are led to an integral equation of the form

$$f(x) = (I + T(k))\mu; \quad x \in \partial D \tag{3.4}$$

where $T(k)$ is a compact integral operator defined on $C^0(\partial D)$. Due to the choice of the fundamental solution $G(\xi;x)$ it can be shown that $(I + T(k))^{-1}$ exists for every positive $k$([16]).

The second approach which we shall find useful in our discussion of scattering problems in a spherically stratified medium is due to Leis ([13]). In this approach a solution of (3.1a)-(3.1c) is sought in the form

$$h(x) = \int_{\partial D} \mu(\xi) \left[\frac{e^{ikR}}{R} + \alpha \frac{\partial}{\partial \nu_\xi} \frac{e^{ikR}}{R}\right] d\omega_\xi \tag{3.5}$$

where $\mu(\xi)$ is a density to be determined, $\alpha$ is a constant such that $\operatorname{Im}\alpha \neq 0$, and $R = |x - \xi|$. Applying the boundary condition (3.1b) we are again led to an integral equation of the form (3.4), but where now $T(k)$ is a singular integral operator. However it can be shown ([11], [13]) that $T(k)$ can be regularized and a solution to (3.4) again exists for all positive $k$. Although this approach to (3.1a)-(3.1c) has the advantage over that of Ursell in having a simple kernel for the integral operator, it has the obvious disadvantage of leading to a singular integral equation for the determination of the unknown density $\mu(\xi)$. However such an equation is amenable to numerical methods (c.f. [9], [12]), and in certain circumstances is computationally more suitable than that of Ursell (c.f. Section IV).

## IV THE SCATTERING OF ACOUSTIC WAVES IN A SPHERICALLY STRATIFIED MEDIUM

In this final section we shall combine the results of Sections II and III to derive constructive methods for solving the acoustic scattering problem modelled by the following system of equations:

$$\Delta_3 u + k^2(1 + B(r))u = 0 \text{ in } \mathbb{R}^3 \backslash D \tag{4.1a}$$

$$u(x) = e^{ikz} + u_s(x) \tag{4.1b}$$

$$\frac{\partial u}{\partial \nu} = 0 \text{ on } \partial D \tag{4.1c}$$

$$\lim_{r\to\infty} r\left(\frac{\partial u_s}{\partial r} - iku_s\right) = 0 . \tag{4.1d}$$

The notation in (4.1a)-(4.1d) is the same as in previous sections except for $u_s(\underset{\sim}{x})$ which denotes the scattered wave where $\underset{\sim}{x} = (x,y,z)$. Note that the scattering of the incoming plane wave $e^{ikz}$ is due to both the inhomogeneous medium described by $B(r) = (\frac{c_o}{c(r)})^2 - 1$ and the "hard" obstacle D, which is again assumed to be bounded, simple connected, and with smooth boundary $\partial D$. A classical approach for studying (4.1a)-(4.1d) is by the method of integral equations, which in general leads to a coupled system of two and three dimensional integral equations which can be solved for k sufficiently small. Since such a formulation is rather unsatisfactory for purposes of numerical approximation, we shall present in this section an alternative approach based on the use of transformation operators. This approach leads to a single two dimensional integral equation over $\partial D$ with the only restriction on k being that needed for the existence of the kernel $E(r,s;k)$ (see Section II). Our primary concern will be for small inhomogeneities $B(r)$ (since in this case the kernels of the transformation operators are readily approximated) and with low and intermediate values of the wave number k. The results of this section are modest extensions of the author's earlier work, viz. in [7] where it was assumed that $B(r)$ had compact support and in [4] where use was made of Gilbert's "method of ascent" since the operator (2.3) was unavailable.

We first consider the case where D is starlike with respect to the origin, $\{\underset{\sim}{x}: |\underset{\sim}{x}| \leq b\} \subset D$, and $0<k<2\gamma b$ where $\gamma$ is defined by (2.5). In this case we look for a solution of (4.1a)-(4.1d) in the form

$$u(r,\theta,\phi) = h(r,\theta,\phi) + \int_r^\infty E(r,s;k)\, h(s,\theta,\phi)ds \qquad (4.2)$$

where, following Ursell's method, $h(\underset{\sim}{x})$ is represented in the form

$$h(\underset{\sim}{x}) = e^{ikz} + \int_{\partial D} \mu(\underset{\sim}{\xi})\, G\, (\underset{\sim}{\xi};\underset{\sim}{x})d\omega_\xi \quad . \qquad (4.3)$$

Substituting (4.3) into (4.2), interchanging orders of integration, and applying the boundary condition (4.1c) leads to an integral equation of the form (3.4) for the determination of the unknown density $\mu(\underset{\sim}{\xi})$, where $T(k)$ is a compact integral operator defined on $C^o(\partial D)$ and $f(\underset{\sim}{x})$ is given by

$$f(\underset{\sim}{x}) = \frac{1}{2\pi}\, \frac{\partial}{\partial\nu}\, [e^{ikz} + \int_r^\infty E(r,s;k)\, e^{iks\cos\theta}ds] \quad . \qquad (4.4)$$

Note that the quantity in brackets in (4.4) is the unique continuation from infinity of $e^{ikz}$ as a solution of (4.1a). For details of the above calculations (in the special case when $B(r)$ has compact support) see [7]. The methods of [7] also yield the following result on the invertibility of the operator $I+T(k)$:

Theorem: Let $0<k<2\gamma b$. Then $(I+T(k))^{-1}$ exists on $C^o(\partial D)$.

We now consider the case when D is no longer starlike with respect to the origin, in particular when $B(r)$ has compact support and D is disjoint from $B = \{\underset{\sim}{x}: B(r) \neq 0,\ r = |\underset{\sim}{x}|\}$. In this case, using the method of Leis, we look for a solution of (4.1a)-(4.1d) in the following form:

$$u(\underset{\sim}{x}) = \begin{cases} e^{ikz} + \sum_{n=0}^{\infty} \sum_{m=-n}^{n} a_{nm} h_n^{(1)}(kr) S_{nm}(\theta,\phi) \\ + \frac{1}{2\pi} \int_{\partial D} \mu(\underset{\sim}{\xi}) [\frac{e^{ikR}}{R} + \alpha \frac{\partial}{\partial \nu_\xi} \frac{e^{ikR}}{R}] d\omega_\xi ; \; \underset{\sim}{x} \in \mathbb{R}^3 \setminus (D \cup B) \\ \sum_{n=0}^{\infty} \sum_{m=-n} b_{nm} u_n(r) S_{nm}(\theta,\phi); \; \underset{\sim}{x} \in B, \end{cases} \tag{4.5}$$

where $h_n^{(1)}(kr)$ denotes a spherical Hankel function, $S_{nm}(\theta,\phi)$ a spherical harmonic, and

$$u_n(r) = j_n(kr) + \int_0^r G(r,s;k) j_n(ks)ds \tag{4.6}$$

where $j_n(kr)$ is a spherical Bessel function. The unknown constants $a_{nm}$, $b_{nm}$, and density $\mu(\underset{\sim}{\xi})$ are to be determined from the conditions a) $u(\underset{\sim}{x}) \in C^2(\mathbb{R}^3 \setminus D)$, and b) $\frac{\partial u}{\partial \nu} = 0$ on $\partial D$ (it is in step a), i.e. determining the constants $a_{nm}$, $b_{nm}$ by continuity conditions across $\partial B$, that makes it advantageous to use the method of Leis to represent $u(\underset{\sim}{x})$ in the homogeneous part of the medium rather than that of Ursell). Such considerations lead to a singular integral equation of the form (3.4) where

$$f(\underset{\sim}{x}) = \frac{\partial}{\partial \nu} [ \sum_{n=0}^{\infty} \frac{c_n}{A_n} h_n^{(1)}(kr) P_n(\cos\theta) - e^{ikz} ] , \tag{4.7}$$

$P_n(\cos\theta)$ denotes Legendre's polynomial, and for $B \subset \{\underset{\sim}{x}: |\underset{\sim}{x}| \leq a\}$

$$c_n = (2n+1) i^n \det \begin{vmatrix} j_n(ka) & u_n(a) \\ j_n'(ka) & u_n'(a) \end{vmatrix}$$

$$A_n = \det \begin{vmatrix} h_n^{(1)}(ka) & u_n(a) \\ h_n'^{(1)}(ka) & u_n'(a) \end{vmatrix} \tag{4.8}$$

with the prime denoting differentiation. The singular part of the operator $T(k)$ is the same as that of Leis for a homogeneous medium, whereas the kernel of the compact part of the operator $T(k)$ is the sum of a weakly singular kernel plus a rapidly convergent series depending on the constants $c_n/A_n$ ([4]). Hence we can conclude that $I+T(k)$ is regularizable. Furthermore, for small inhomogeneities $B(r)$, $u_n(r)$ can be accurately computed from (4.6) as a small perturbation of $j_n(kr)$ and hence from (4.7), (4.8), $f(\underset{\sim}{x})$ (as well as the kernel of the compact part of the operator $T(k)$) can be easily approximated. Finally, it is possible to prove the following Theorem ([4]):

<u>Theorem</u>: For every positive k, $(I+T(k))^{-1}$ exists on $C^0(\partial D)$.

## REFERENCES

1. Z. S. Agranovich and V. A. Marchenko, The Inverse Problem of Scattering Theory, Gordon and Breach, London, 1963.

2. B. L. J. Braaksma, A singular Cauchy problem and generalized translations, in International Conference on Differential Equations, Academic Press, New York, 1975, 40-52.

3. L. M. Brekhovskikh, Waves in Layered Media, Academic Press, New York, 1965.

4. D. Colton, The scattering of acoustic waves by a spherically stratified medium and an obstacle, SIAM J. Math. Anal., to appear.

5. D. Colton, G. Hsiao, and R. Kress, in preparation.

6. D. Colton and R. Kress, The construction of solutions to acoustic scattering problems in a spherically stratified medium, Quart J. Mech. Appl. Math., to appear.

7. D. Colton and W. Wendland, Constructive methods for solving the exterior Neumann problem for the reduced wave equation in a spherically symmetric medium, Proc. Roy. Soc. Edin. 75A(1976), 97-107.

8. R. P. Gilbert, Constructive Methods for Elliptic Equations, Springer-Verlag Lecture Note Series Vol. 365, Springer-Verlag, Berlin, 1974.

9. V. V. Ivanov, The Theory of Approximate Methods and Their Application to the Numerical Solution of Singular Integral Equations, Noordhoff International Publishing, Leyden, 1976.

10. D. S. Jones, Integral equations for the exterior acoustic problem, Quart. J. Mech. Appl. Math. 27(1974), 129-142.

11. R. Kress and G. F. Roach, On mixed boundary value problems for the Helmholtz equation, Proc. Roy. Soc. Edin. 77A (1977), 65-77.

12. R. Kussmaul, Ein Numerisches Verfahren zur Lösung des Neumannschen Aussenraumproblems für die Helmholtzche Schwingungsgleichung, Computing 4(1969), 246-273.

13. R. Leis, Vorlesungen über Partielle Differentialgleichungen Zweiter Ordrung, BI Mannheim, Mannheim, 1967.

14. B. Levitan, Generalized Translation Operators and Some of their Applications, Israel Program for Scientific Translations, Jerusalem, 1964.

15. G. O. Olaofe and S. Levine, Electromagnetic scattering by a spherically symmetric inhomogeneous particle, in Electromagnetic Scattering, Gordon and Breach, London, 1967, 237-292.

16. F. Ursell, On the exterior problems of acoustics, Proc. Camb. Phil. Soc. 74(1973), 117-125.

*Differential Equations and Applications*
*W. Eckhaus and E.M. de Jager (eds.)*
*©North-Holland Publishing Company (1978)*

# DERIVATION OF FORMULAS RELEVANT TO NEUTRON TRANSPORT IN MEDIA WITH ANISOTROPIC SCATTERING.

R.J. HANGELBROEK
Catholic University
Nijmegen, The Netherlands

## INTRODUCTION

We discuss some problems related to the solution of the equation

$$(1)\qquad \mu\frac{\partial\psi}{\partial x}(x,\mu) = -\psi(x,\mu) + \sum_{j=0}^{n} b_j \int_{-1}^{+1} \psi(x,\mu')p_j(\mu')d\mu' p_j(\mu).$$

Here x ranges over some interval $J \subset \mathbb{R}$ and $\mu \in I = [-1,+1]$. The (n+1) coefficients $b_j$ are given real constants satisfying the inequalities

$$(2)\qquad |b_j| \leq b_0 < 1.$$

The $p_j$ denote the Legendre polynomials normalized so that $\int_{-1}^{+1} \{p_j(\mu)\}^2 d\mu = 1$.

Equation (1) describes the time-independent transport of neutrons in homogeneous media with anisotropic scattering and plane geometry. The unknown function $\psi$ is the neutron density in phase space (x is a position coordinate and $\mu$ the cosine of the angle between velocity and x-direction) and, hence, a non-negative real valued function. The sum of integrals in equation (1) originates from an approximation of some integral operator working on $\psi$. The assumption $b_0 < 1$ in (2) means that we are dealing with an absorbing medium.

Typical boundary value problems are:

I. The finite slab problem: J = (0,1) and the conditions

$$\psi(0,\mu) = g_+(\mu), \quad \mu \in I_+ = [0,1];$$

$$\psi(1,\mu) = g_-(\mu), \quad \mu \in I_- = [-1,0]$$

and

II. The half-space problem: $J = (0,\infty)$ and the conditions

$$\psi(0,\mu) = g_+(\mu), \quad \mu \in I_+ ;$$

$$\lim_{x\to\infty} \psi(x,\mu) = 0, \quad \mu \in I.$$

The change of sign of the coefficient of $\frac{\partial\psi}{\partial x}$ in (1) as $\mu$ goes through zero as well as the nature of the boundary conditions reflect the physical situation of particles moving in the positive and negative x-directions.

Equation (1) and the associated boundary value problems have been attacked by many authors using widely varying techniques. A particularly sucessful method, based on the idea of separation of variables, was presented by K.M. Case in 1960 in a paper on the isotropic case (i.e., n=0) ([1], see also [2]). This success led to two independent, functional analytic approaches by Larsen and Habetler, [3], and the author of this paper, [4], both also for n = 0. From a point of view of analysis the possibility to extend Case's method as well as the two functional analytic approaches to the anisotropic case was rather evident. However, the actual construction of more or less explicit solutions had to suffer apparently from computational complications caused by the larger number of coefficients $b_j$ ([5], [6]). The purpose of this paper is to show how contour integral techniques can be utilized to obtain the relevant explicit formulas for both the isotropic and anisotropic (with finite n) cases in the same, rather simple way.

In order to indicate which formulas we have in mind we will give now an outline of our approach.

We do consider equation (1) as a first order ordinary differential equation for functions $\psi$ of a real variable x with values in the Hilbert space $H = L^2(I)$. To this end we define two bounded linear operators T and A in H : T is the operator of multiplication by the independent variable $\mu$ and A is defined by

$$(3) \qquad A\phi = \phi - \sum_{j=0}^{n} b_j(\phi,p_j)p_j, \qquad \phi \in H,$$

where ( , ) denotes the usual inner product in $L^2(I)$. The operator A has a bounded inverse. Hence, we can write equation (1) in the form

$$(4) \qquad A^{-1}T\frac{d\psi}{dx}(x) = -\psi(x), \qquad x \in J,$$

where the bounded operator $A^{-1}T$ has an unbounded inverse.
The space $H = L^2(I)$ can be considered as the direct sum of the spaces $H_+ = L^2(I_+)$ and $H_- = L^2(I_-)$:

(5) $$H = H_+ \oplus H_-.$$

We denote the projections onto $H_{\pm}$ along $H_{\mp}$ by $P_{\pm}$. The boundary conditions for the problems I and II will be taken as

$$\text{I.} \quad P_+\psi(0) = g_+, \; g_+ \in H_+; \; P_-\psi(1) = g_-, \; g_- \in H_-$$

and

$$\text{II.} \quad P_+\psi(0) = g_+, \; g_+ \in H_+; \; \lim_{x\to 0} \psi(x) = 0 \text{ in } H.$$

The spaces $H_+$ and $H_-$ considered as subspaces of H are the maximal invariant subspaces in which T is positive and negative, respectively. The space H can be decomposed also as

(6) $$H = H_p \oplus H_m$$

where $H_p$ and $H_m$ are the maximal invariant subspaces in which the operator $A^{-1}T$ is positive and negative, respectively. Let $P_p$ and $P_m$ denote the projections onto $H_p$ and $H_m$ along $H_m$ and $H_p$, respectively. Then we obtain as solution to problem I:

(7) $$\psi(x) = e^{-xT^{-1}A} P_p h + e^{(1-x)T^{-1}A} P_m h, \quad 0 < x < 1$$

where $h \in H$ has to satisfy

(8) $$(P_+P_p + P_-P_m + P_+e^{T^{-1}A}P_m + P_-e^{-T^{-1}A}P_p)h = g_+ + g_-,$$

and to problem II:

(9) $$\psi(x) = e^{-xT^{-1}A}h_p, \quad x > 0,$$

where $h_p \in H_p$ has to satisfy

(10) $$P_+h_p = g_+.$$

By using the spectral theorem applied to the operator $A^{-1}T$ (see section 2) one may justify the steps taken in the above given outline. In particular, the definitions of the operators $P_p$, $P_m$, $e^{-xT^{-1}A}P_p$ and $e^{(1-x)T^{-1}A}P_m$ which feature in the equations (7) and (9) can be based upon this theorem. However, the main purpose of this paper is not to present such a justification, but to show (in section 3) how the results of the spectral theorem can be substantiated. This substantiation is needed if one wants to put the equations (7) and (9) in a concrete form. In a forthcoming paper [7], we hope to show how the results presented in section 3 can be employed to solve the so-called half range problem which we presented above in the form of equation (10).

## 2. The operator $A^{-1}T$.

In this section we present the main properties of the operator $A^{-1}T$ introduced in section 1. We start with the remark that A has the eigenvalues $(1-b_j)$, $j = 0,\ldots,n$ and 1 with the set $\{p_j\}_{j=0}^{\infty}$ as a complete set of orthonormal eigenvectors. On the strength of the assumption (2) we can estimate

$$(11)\qquad (1-b_0)(\phi,\phi) \leq (A\phi,\phi) \leq (1+b_0)(\phi,\phi), \qquad \phi \in H.$$

Hence, A is positive definite, since $1-b_0 > 0$. The inverse of A can be readily found to be

$$(12)\qquad A^{-1}\phi = \phi + \sum_{j=0}^{n} \gamma_j(\phi,p_j)p_j, \quad \phi \in H, \quad \gamma_j = \frac{b_j}{1-b_j}.$$

Now, the operator $A^{-1}T$ is a well defined, bounded operator. It is not self adjoint in H relative to the inner product ( , ), but it is self adjoint relative to the equivalent inner product $(\ ,\ )_A$ defined by

$$(\phi,\psi)_A = (A\phi,\psi) \qquad \phi,\psi \in H.$$

An important property of $A^{-1}T$ is that for $\phi$ a polynomial function on I of degree j $A^{-1}T\phi$ will be a polynomial function of degree j+1. The proof can be given directly but we like to show the property via the recurrence relation for the Legendre polynomials

$$(2j+1)TP_j = (j+1)P_{j+2}+jP_{j-1}, \; j = 0,1,\ldots \; (P_{-1} \equiv 0, \; P_0 \equiv 1).$$

Application of the operator $A^{-1}$ yields

$$(13) \qquad (2j+1)A^{-1}TP_j = \frac{j+1}{1-b_{j+1}} P_{j+1} + \frac{j}{1-b_{j-1}} P_{j-1}$$

with $b_j = 0$ for $j = n+1, n+2, \ldots$

A consequence of the property is that if $e \in H$ is the function defined by $e(\mu) = 1$, $\mu \in I$ and if $\tilde{\phi}$ is a polynomial of degree j then $\tilde{\phi}(A^{-1}T)e$ is a polynomial function on I of degree j+1. Hence, e is a cyclic vector in H for the algebra of polynomials of the operator $A^{-1}T$.

We are now able to formulate the spectral theorem applied to the operator $A^{-1}T$:

<u>Spectral theorem</u>. There exist a finite, positive Borel measure $\sigma$ on the spectrum N of the operator $A^{-1}T$ and a linear transformation F that maps H onto the space $L^2(N,\sigma)$ of complex valued functions which are square integrable with respect to $\sigma$, such that

(i) $(\phi,\psi)_A = \int_N (F\phi)(\nu)\overline{(F\psi)(\nu)}\, d\sigma(\nu), \qquad \phi,\psi \in H$

(ii) $FA^{-1}T\phi = T_N F\phi, \qquad \phi \in H,$

where $T_N$ is defined by $(T_N\tilde{\phi})(\nu) = \nu\tilde{\phi}(\nu)$, $\tilde{\phi} \in L^2(N,\sigma)$, $\nu \in N$

(iii) F and $\sigma$ are uniquely determined by the condition $Fe = \tilde{e}$ with $\tilde{e}(\nu) = 1$, $\nu \in N$.

We will denote the inner product defined by the right member of (i) by $(\ ,\ )_N$. We remark that as a consequence of (ii) and (iii) one obtains that for $\tilde{\phi}$ a polynomial $F\tilde{\phi}(A^{-1}T)e$ will be the same polynomial $\tilde{\phi}$ but now considered as a polynomial function on N, whence $F\tilde{\phi}(A^{-1}T)e = \tilde{\phi}$.

## 3. The spectrum N, the measure $\sigma$ and the transforms F, $F^{-1}$.

In this final section we determine the spectrum N, the measure $\sigma$ and the transforms F and $F^{-1}$ which feature in the spectral theorem for the operator $A^{-1}T$ as given in section 2. It is convenient to remark firstly that the transforms F and $F^{-1}$ essentially will be known when one knows how the complete set of Legendre polynomials in H is transformed by F. But a recurrence relation for these polynomial functions $FP_j$ is obtained directly by application of the transformation F to the relation (13):

$$(14)\quad (2j+1)T_N H_j = \frac{j+1}{1-b_{j+1}} H_{j+1} + \frac{j}{1-b_{j-1}} H_{j-1},\ j = 0,1,2,\ldots\ (H_{-1} = 0,\ H_0 = \tilde{e}),$$

where we have denoted $H_j = FP_j$. Hence, we may consider the F-transforms of the ordinary as well as the normalized Legendre polynomials as known. We write also $h_j = Fp_j$. Formula (23) will show that for large classes of functions on I we actually do need to know only the first n+1 polynomial functions $h_j$ in order to represent the action of the transformation F.

In the remaining derivations we will finally use of A its most salient property, viz., that it differs from the identity operator by a finite dimensional operator. A direct consequence is that the operator $A^{-1}T$ is obtained by perturbation of the multiplication operator T by a finite rank operator since we may write (using relation (12))

$$(15)\qquad A^{-1}T = (E+B)T = T+BT$$

where E denotes the identity operator and B is defined by

$$B\phi = \sum_{j=0}^{n} \gamma_j(\phi,p_j)p_j, \qquad \phi \in H.$$

It is not remarkable that in order to obtain results for the perturbed operator $A^{-1}T$ from the properties of the unperturbed operator T we will have to study the operator

$$(16)\qquad (\lambda-A^{-1}T)(\lambda-T)^{-1} = E-BT(\lambda-T)^{-1}, \qquad \lambda \notin I$$

but it turns out that for the following derivations we need only to know how the latter operator and its inverse work upon the function e:

$$(17)\qquad (\lambda-A^{-1}T)(\lambda-T)^{-1}e = e + \sum_{j=0}^{n} \gamma_j \int_{-1}^{+1} \frac{\mu}{\mu-\lambda}\, p_j(\mu)d\mu\ p_j, \qquad \lambda \notin I$$

and

$$(18)\qquad (\lambda-T)(\lambda-A^{-1}T)^{-1}e = e - \sum_{j=0}^{n} \gamma_j \int_N \frac{\nu}{\nu-\lambda}\, h_j(\nu)d\sigma(\nu)p_j, \quad \lambda \notin N$$

where the latter relation has been obtained by the use of statement (i) of the spectral theorem given in section 2.

Determination of the spectrum N of $A^{-1}T$. According to a theorem on degenerate perturbations (e.g., see [8], Th.IV 6.2) the spectrum N of $A^{-1}T$ will consist out of the interval $I = [-1,+1]$, which is the spectrum of T, and the zeros of the W-A determinant of the pair of operators T, $A^{-1}T$. This determinant can be defined by

$$(19) \qquad \det\{(\lambda-A^{-1}T)(\lambda-T)^{-1}\} = \det\{(E-BT(\lambda-T)^{-1})|P_n(I)\}, \quad \lambda \notin I$$

where the right member indicates that the determinant has to be taken of the operator $E-BT(\lambda-T)^{-1}$ restricted to the range of the perturbation BT, which consists of the subspace $P_n(I)$ of polynomial functions on I of degree at most n. In order to evaluate the determinant one may set up the matrix representation of the operator in $P_n(I)$ relative to, e.g., the Legendre polynomials, but it is easier to remark that the two sets $\{e,(\lambda-T)p_0,\dots,(\lambda-T)p_{n-1}\}$ and $\{e,(\lambda-A^{-1}T)p_0,\dots,(\lambda-A^{-1}T)p_{n-1}\}$ constitute bases of the subspace $P_n(I)$. Then one may use the definition of a determinant as an a-symmetric, multilinear form to prove in a rahter simple way that the coefficient $\Lambda(\lambda)$ of e in the expansion

$$(20) \qquad (\lambda-A^{-1}T)(\lambda-T)^{-1}e = \Lambda(\lambda)e + \sum_{i=0}^{n-1} \alpha_i(\lambda)(\lambda-A^{-1}T)p_i, \qquad \lambda \notin I$$

has to be equal to the sought determinant up to a nonzero constant factor. In fact, $\Lambda(\lambda) = \prod_{j=1}^{n} (1-b_j) \det\{(\lambda-A^{-1}T)(\lambda-T)^{-1}\}$. The coefficient $\Lambda(\lambda)$ can be obtained as follows. We identify the right members of (17) and (20) and apply the transform F to obtain

$$\Lambda(\lambda) + \sum_{i=0}^{n-1} \alpha_i(\lambda)(\lambda-\nu)h_i(\nu) = 1 + \sum_{j=0}^{n} \gamma_j\, h_j(\nu) \int_{-1}^{+1} \frac{\mu}{\mu-\lambda}\, p_j(\mu)d\mu,$$

where we have taken the $\nu$-values of the polynomial functions on either side. Putting $\nu = \lambda$ we obtain

$$(21) \qquad \Lambda(\lambda) = 1 + \sum_{j=0}^{n} \gamma_j\, h_j(\lambda) \int_{-1}^{+1} \frac{\mu}{\mu-\lambda}\, p_j(\mu)d\mu, \qquad \lambda \notin I.$$

The function $\Lambda$ multiplied by $(1-b_0)$ equals a function, denoted by the same symbol, which in transport theory is called the dispersion function. Using the principle of the argument one can show that it has $2n_0$ zeros, $\{\pm\nu_j\}_{j=1}^{n_0}$, outside the interval I with $n_0 \le n+1$, ([5]).

Determination of the measure $\sigma$. In a way similar to the one we went to obtain relation (21) we may use the equation (obtained from (20))

$$(\lambda-T)(\lambda-A^{-1}T)^{-1}e = \frac{1}{\Lambda(\lambda)} e - \sum_{i=0}^{n-1} \frac{\alpha_i(\lambda)}{\Lambda(\lambda)} (\lambda-T)p_i, \qquad \lambda \notin N$$

and equation (18) to find

$$(22) \qquad \frac{1}{\Lambda(\lambda)} = 1 - \sum_{j=0}^{n} \gamma_j p_j(\lambda) \int_N \frac{\nu}{\nu-\lambda} h_j(\nu)d\sigma(\nu), \qquad \lambda \notin N.$$

The latter equation contains the measure $\sigma$ as an unknown quantity. After some analysis of, essentially, the equations (21) and (22) one is able to determine $\sigma$: at the isolated points $\nu_j$ of N in terms of the residues of $1/\Lambda(\lambda)$ at those points and on the interval [-1,+1] in the form $d\sigma(\nu) = d\nu/\Lambda^+(\nu)\Lambda^-(\nu)$, where $\Lambda^{\pm}$ denote the boundary values of the dispersion function $\Lambda$, (see [9]).

Determination of the transforms F and $F^{-1}$. Let $\phi$ be a polynomial function on I then one can use the Dunford-Taylor integral representation of the operator $\phi(T)$ to write

$$\phi = \phi(T)e = \frac{1}{2\pi i} \int_\Gamma \phi(\lambda)(\lambda-T)^{-1}ed\lambda,$$

where $\Gamma$ is any simple, positively oriented contour enclosing the set N. Hence,

$$F\phi = \frac{1}{2\pi i} \int_\Gamma \phi(\lambda)\ F(\lambda-T)^{-1}ed\lambda.$$

The factor $F(\lambda-T)^{-1}e$ can be obtained directly from (17). If one takes the values at any point $\nu \in N$ and interchanges the order of integration one derives

$$(23) \qquad (F\phi)(\nu) = \phi(\nu) - \sum_{j=0}^{n} \gamma_j h_j(\nu) \int_{-1}^{+1} \mu p_j(\mu)\frac{\phi(\mu)-\phi(\nu)}{\mu - \nu}d\mu, \qquad \nu \in N.$$

In order to obtain an expression for $F^{-1}$ when applied to polynomial functions $\tilde{\phi}$ on N, we use the integral representation for the operator $\tilde{\phi}(A^{-1}T)$ to write (see the remark at the end of section 2)

$$F^{-1}\tilde{\phi} = \tilde{\phi}(A^{-1}T)e = \frac{1}{2\pi i}\int_{\Gamma} \tilde{\phi}(\lambda)(\lambda-A^{-1}T)^{-1}e\,d\lambda,$$

where $\Gamma$ is again a contour enclosing N. The factor $(\lambda-A^{-1}T)^{-1}e$ is taken from equation (18) and a procedure similar to the one used above then yields

$$(F^{-1}\tilde{\phi})(\mu) = \tilde{\phi}(\mu) + \sum_{j=0}^{n} \gamma_j p_j(\mu) \int_N \nu h_j(\nu)\,\frac{\tilde{\phi}(\nu)-\tilde{\phi}(\mu)}{\nu-\mu}\,d\sigma(\nu), \quad \mu \in I. \tag{24}$$

The formulas (23) and (24) have been derived for polynomial functions. However, they can be extended to larger classes of functions, e.g., functions which are Hölder continuous on I and N, respectively, (see [4], where such an extension was carried out in the isotropic case $n = 0$).

REFERENCES


1. K.M. CASE, Elementary solutions of the transport equation and their application. Ann. Phys., 9 (1960) p.1.
2. K.M. CASE and P.F. ZWEIFEL, Linear transport theory. Addison-Wesley Reading, Mass., 1967.
3. E.W. LARSEN and G.J. HABETLER, A functional-analytic derivation of Case's full and half range formulas. Comm. Pure Appl. Math., 26 (1973) p.525.
4. R.J. HANGELBROEK, Linear analysis and solutions of neutron transport problems. Transp. Th. Stat. Phys., 5 (1976) p.1. (Revision of thesis: A functional analytic approach to the linear transport equation. Groningen Univ. 1973.)
5. J.R. MIKA, Neutron transport with anisotropic scattering. Nuc. Sci. Eng., 11 (1961) p.415.
6. E.W. LARSEN, A functional-analytic approach to the steady, one-speed neutron transport equation with anisotropic scattering. Comm. Pure Appl. Math. 27 (1974) p.523.
7. R.J. HANGELBROEK, The linear transport equation in a medium with anisotropic scattering (Degenerate scattering function), in preparation.
8. T. KATO, Perturbation theory for linear operators. Springer, Berlin, 1966.
9. R.J. HANGELBROEK, On the derivation of some formulas in linear transport theory for media with anisotropic scattering. Nijmegen, Cath. Univ., Math. Dept. Report 7720.



ACKNOWLEDGEMENT

This study was partially supported by the U.S. Energy Research and Development Administration during the author's stay at Argonne National Laboratory, Argonne, Illinois 60439.

*Differential Equations and Applications*
*W. Eckhaus and E.M. de Jager (eds.)*
*©North-Holland Publishing Company (1978)*

# THE GENERALIZED GREEN'S FUNCTION FOR REGULAR ORDINARY DIFFERENTIAL SUBSPACES IN $L^2[a,b] \oplus L^2[a,b]$

Aalt Dijksma

## INTRODUCTION

Let L be a regular ordinary differential expression of order n acting on $m \times 1$ matrix-valued functions defined on the finite closed interval $[a,b] \subset \mathbb{R}$. Let $H = L^2[a,b]$ be the Hilbertspace of $m \times 1$ matrix-valued functions f on [a,b] for which $\int_b^a f^* f < \infty$. Associate with L a differential operator T in $H$. For example, let

$$T = \{\{f, Lf\} \mid f \in C^{n-1}[a,b],\ f^{(n-1)} \in AC[a,b],\ Lf \in H, \text{ and}$$
$$\int_a^b d\mu_j^* f^{(j-1)} = 0,\ j = 1,2,\ldots,n\},$$

where $\mu_j$ is an $m \times p_j$ matrix-valued function of bounded variation on [a,b]. Note that we have identified T with its graph in $H^2 = H \oplus H$.

Clearly, the inverse $T^{-1}$ of T, defined by

$$T^{-1} = \{\{g,f\} \mid \{f,g\} \in T\},$$

need not be (the graph of) an operator. However, there exists an operator $G$, called the <u>generalized inverse</u> of T, which is bounded and defined on all of $H$ and replaces $T^{-1}$ in the sense that if $Tf = g$ then $TGg = g$ also. It turns out that $G$ is an integral operator with kernel $G \in L^2([a,b] \times [a,b])$. This function G is called the <u>generalized Green's function</u>. It has been studied in a large number of papers of which we mention Reid [9], [10], Loud [7], [8], Bradley [1], Chitwood [3], Brown [2], and Locker [6]. In these papers more references can be found.

The adjoint $T^*$ of T, defined by

$$T^* = \{\{h,k\} \mid \text{for all } \{f,g\} \in T,\ (g,h) - (f,k) = 0\},$$

plays an important role in the construction of the kernel G. Furthermore the adjoint $G^*$ of $G$ is the generalized inverse of $T^*$. In general $T^*$ as given in the example above need not be (the graph of) an operator but it is always a closed linear manifold in $H^2$. It is not (the graph of) an operator when the side conditions such as $\int_a^b d\mu_j^* f^{(j-1)} = 0$ are so restrictive that T is no longer densely defined. Subspaces (= closed linear manifolds) and their adjoints are discussed in detail in Coddington and Dijksma [4]. The theory developed in that paper gives rise to non-densely defined, multivalued integral-ordinary differential-boundary operators, in short, <u>ordinary differential subspaces</u> in $H^2$.

For such subspaces we will construct the generalized Green's function and discuss its properties. Let us briefly summarize the contents of the following sections. In Section 1 we define the notion of subspaces and adjoint subspaces, and prove, under hypotheses almost immediately applicable to differential subspaces, a representation theorem. In Section 2 we give the definition of the generalized inverse of a subspace and list some of its properties. As an example of a generalized inverse we give an explicit formula for the so called <u>Moore-Penrose</u> inverse of an arbitrary square matrix with coefficients in $\mathbb{C}$. In Section 3 we define and discuss regular ordinary differential subspaces. In Section 4 we show that the generalized inverse of a differential subspace is an integral operator. We give a formula for the kernel of this operator and a characterization of the kernel in terms of jumps on the diagonal of the square $[a,b] \times [a,b]$, differentiability and boundary conditions. In Section 5 we give several examples.

We remark that our results can easily be carried over to the case of constant $n \times m$ matrices and to the case of regular ordinary differential subspaces in $L^p[a,b] \oplus L^q[a,b]$, $1/p + 1/q = 1$, or in $C[a,b] \oplus C[a,b]$.

## 1. SUBSPACES AND ADJOINT SUBSPACES

Let $H$ be a Hilbertspace with innerproduct ( , ) and $H^2 = H \oplus H$, considered as a Hilbertspace. Let T be a linear manifold in $H^2$. The <u>domain</u> $D(T)$, the <u>range</u> $R(T)$ and the <u>null space</u> $\nu(T)$ of T are defined by

$$D(T) = \{f \in H \mid \text{for some } g \in H,\ \{f,g\} \in T\},$$
$$R(T) = \{g \in H \mid \text{for some } f \in H,\ \{f,g\} \in T\},$$
$$\nu(T) = \{f \in H \mid \{f,0\} \in T\}.$$

For $f \in D(T)$ we let

$$T(f) = \{g \in H \mid \{f,g\} \in T\}.$$

T is the graph of a linear operator if and only if $T(0) = \{0\}$. If T is the graph of a linear operator we denote this operator by T also and we write $T(f) = Tf$, $f \in D(T)$. If T is a linear operator we shall frequently identify

T with its graph in $H^2$.

We consider linear manifolds as linear relations. If T, S are linear manifolds in $H^2$ and $\alpha \in \mathbb{C}$ we define

$$\alpha T = \{\{f,\alpha g\} \in H^2 \mid \{f,g\} \in T\}$$
$$T^{-1} = \{\{g,f\} \in H^2 \mid \{f,g\} \in T\}$$
$$T + S = \{\{f,g+h\} \in H^2 \mid \{f,g\} \in T, \{f,h\} \in S\},$$
$$TS = \{\{f,h\} \in H^2 \mid \text{for some } g \in H, \{f,g\} \in S, \{g,h\} \in T\}.$$

The product TS is associative, i.e., if R is a linear manifold in $H^2$ also, then (TS) R = T(SR) and we write (TS)R = TSR.

The algebraic sum $T \dot{+} S$ of two linear manifolds T,S in $H^2$ is defined by

$$T \dot{+} S = \{\{f + h, g + k\} \mid \{f,g\} \in T, \{h,k\} \in S\}.$$

This sum is called a direct sum if $T \cap S = \{\{0,0\}\}$.

The multivalued part $T_\infty$ of a linear manifold T in $H^2$ is defined by

$$T_\infty = \{\{0,g\} \in H^2 \mid \{0,g\} \in T\},$$

and a linear manifold $T_s$ in $H^2$ is called an algebraic operator part of T if

$$T = T_s \dot{+} T_\infty, \text{ direct sum.}$$

Clearly an algebraic operator part is a linear operator.

A subspace T in $H^2$ is a closed linear manifold T in $H^2$. A closed algebraic operator part is called an operator part. Every subspace T in $H^2$ has an operator part. Take, for example, $T_s = T \ominus T_\infty$, the orthogonal complement of $T_\infty$ in T.

Let T be a linear manifold in $H^2$. Its adjoint $T^*$ is defined by

$$T^* = \{\{h,k\} \in H^2 \mid \text{for all } \{f,g\} \in T, (g,h) - (f,k) = 0\}.$$

Note that $T^*$ is the orthogonal complement of T with respect to the non-degenerate semibilinear form $\langle\ ,\ \rangle$ on $H^2 \times H^2$ defined by

$$\langle\{f,g\}, \{h,k\}\rangle = (g,h) - (f,k), \{f,g\}, \{h,k\} \in H^2.$$

Clearly, $T^*$ is a subspace in $H^2$ and, if T is a subspace also, $T^{**} = T$. Also, $(T^{-1})^* = (T^*)^{-1}$, $(T \dot{+} S)^* = T^* \cap S^*$ and $(TS)^* = S^*T^*$ for linear manifolds T and S in $H^2$.

Let T be a subspace in $H^2$. Then

$$\nu(T) = R(T^*)^\perp, \quad \nu(T^*) = R(T)^\perp,$$
$$T(0) = D(T^*)^\perp, \quad T^*(0) = D(T)^\perp,$$

where $\perp$ denotes the orthogonal complement in $H$, and $R(T)$ is closed in $H$

if and only if $R(T^*)$ is closed in $H$. Furthermore if B is a finite dimensional subspace in $H^2$, then

$$(T \cap B)^* = T^* \dotplus B^*.$$

Proofs of these and related results can be found in [4; Section 2].

In order to state and prove the following representation theorem we introduce the following notations. To indicate that a matrix M has p rows and q columns we write $M(p \times q)$. The $p \times q$ zero matrix and the $p \times p$ identity matrix are denoted by $O_p^q$ and $I_p$. By $F(1 \times p) \in H$, ${}^1\theta(1 \times p)$ is a basis for $T(0)$ or $\{\sigma,\tau\}$ $(1 \times p)$ is linearly independent we mean that the p columns of these matrices have these properties. If ${}^1\theta(1 \times p) \in H$ and ${}^1\theta = ({}^1\theta_1, \ldots, {}^1\theta_p)$ then $\{0, {}^1\theta\} = (\{0,{}^1\theta_1\}, \ldots, \{0,{}^1\theta_p\})$. For F $(1 \times p)$, $G(1 \times q) \in H$ we define the "matrixinnerproduct" $(F,G)$ to be the $q \times p$ matrix whose i, j-th element is

$$(F,G)_{ij} = (F_j,G_i).$$

If $C(p \times r)$, $D(q \times s)$ are constant matrices, then

$$(FC, G) = (F,G)C \quad , \quad (F,GD) = D^*(F,G).$$

If $\{f,g\}$ $(1 \times p)$, $\{h,k\}$ $(1 \times q) \in H^2$, then it follows that

$$\langle\{f,g\}, \{h,k\}\rangle = (g,h) - (f,k)$$

is a $q \times p$ matrix. We denote by $(F:G)$ the matrix whose columns are obtained by placing the columns of G next to those of F in the order indicated.

The following theorem will be used in Section 4.

Theorem 1.1. Let $A_o \subset A_1$ be subspaces in $H^2$ such that

$$t = \dim A_1(0) < \infty, \; t^+ = \dim A_o^*(0) < \infty.$$

Assume that $A_1$ has an operator part R which is bounded and defined on all of $H$ and that $R^*$ is an operator part of $A_o^*$.

Let A be a subspace in $H^2$ satisfying

(i) $A_o \subset A \subset A_1$, $\dim A(0) = t_1$, $\dim A^*(0) = t_1^+$.

Then

$$\text{(ii)} \begin{cases} A_1^* \subset A^* \subset A_o^*, \\ \dim(A/A_o) = t_1 + t_2^+, \; \dim(A^*/A_1^*) = t_2 + t_1^+, \end{cases}$$

where $t_1 + t_2 = t$, $t_1^+ + t_2^+ = t^+$. Let

(iii) $\begin{cases} A_1(0) = \Theta_1 + \Theta_2, & A_o^*(0) = \Theta_1^+ + \Theta_2^+, \text{ direct sums,} \\ \dim \Theta_j = t_j, & \dim \Theta_j^+ = t_j^+, \quad j = 1,2, \end{cases}$

be decompositions of $A_1(0)$, $A_o^*(0)$ into linear manifolds in $H$ such that

(iv) $A(0) = \Theta_1$, $A^*(0) = \Theta_1^+$.

Let ${}^j\theta$, ${}^j\theta^+$ be bases for $\Theta_j$, $\Theta_j^+$, $j = 1,2$, respectively. Then there exists a constant $t_2 \times t_2^+$ matrix H such that

(v) $A = \{\{\alpha, R\alpha + {}^2\theta H(\alpha, {}^2\theta^+) + {}^1\theta c\} \mid \alpha \in H, (\alpha, {}^1\theta^+) = 0_{t_1^+}^1$, and

$c(t_1 \times 1) \in \mathbb{C}^{t_1}$ is arbitrary$\}$,

and

(vi) $A^* = \{\{\alpha^+, R^*\alpha^+ + {}^2\theta^+ H^*(\alpha^+, {}^2\theta) + {}^1\theta^+ c^+\} \mid \alpha^+ \in H, (\alpha^+, {}^1\theta) = 0_{t_1}^1$, and

$c^+ (t_1^+ \times 1) \in \mathbb{C}_1^{t_1^+}$ is arbitrary$\}$.

Conversely, if (iii) represent decompositions of $A_1(0)$, $A_o^*(0)$ and ${}^j\theta$, ${}^j\theta^+$ are bases for $\Theta_j$, $\Theta_j^+$, $j = 1,2$, respectively, and A is defined by (v), where H is any constant $t_2 \times t_2^+$ matrix, then A is a subspace in $H^2$ satisfying (i), (ii), (iv) and (vi).

Proof. We will first prove that (i) implies (ii). Clearly the inclusions in (i) imply the inclusions in (ii). Since

(1.1) $A_1 = R \dot{+} (A_1)_\infty$, $A_o^* = R^* \dot{+} (A_o^*)_\infty$, direct sums,

we have that

(1.2) $A_o = (A_o^*)^* = R \cap ((A_o^*)_\infty)^*$, $A_1^* = R^* \cap ((A_1)_\infty)^*$.

It follows that $A_o$ and $A_1^*$ are operators with closed domains. Hence

$$D(A_o) = (A_o^*(0))^\perp, \quad D(A_1^*) = (A_1(0))^\perp.$$

Also it follows that

(1.3) $\dim(A_1/A_o) = \dim(A_o^*/A_1^*) = t + t^+ < \infty$,

which implies that there exist finite dimensional subspaces D, $D^+$ in $H^2$ such that

$$A = A_o \dot{+} D \dot{+} (A)_\infty, \quad A^* = A_1^* \dot{+} D^+ \dot{+} (A^*)_\infty, \text{ direct sums.}$$

It follows that A and A* have closed domains also. Hence

$$\mathcal{D}(A) = (A^*(0))^{\perp}, \quad \mathcal{D}(A^*) = (A(0))^{\perp}.$$

It is easy to see that D and $D^+$ are operators and thus we have that

$$\dim D = \dim \mathcal{D}(D) = \dim(\mathcal{D}(A)/\mathcal{D}(A_0)) = t^+ - t_1^+ = t_2^+,$$
$$\dim D^+ = \dim \mathcal{D}(D^+) = \dim(\mathcal{D}(A^*)/\mathcal{D}(A_1^*)) = t - t_1 = t_2 .$$

Hence

$$\dim(A/A_0) = \dim D + \dim (A)_\infty = t_2^+ + t_1,$$
$$\dim(A^*/A^*_1) = \dim D^+ + \dim (A^*)_\infty = t_2 + t_1^+.$$

Thus, indeed, (i) implies (ii).

Let $\{\sigma,\tau\}$ $(1 \times t_2^+)$, $\{\sigma^+,\tau^+\}$ $(1 \times t_2)$ be bases for D, $D^+$ respectively. Then it follows from (1.1) that there exist $\{u,Ru\}(1 \times t_2^+) \in R$, $\{u^+,R^*u^+\}(1 \times t_2) \in R^*$ and constant matrices $C_1(t_1 \times t_2^+)$, $C_2(t_2 \times t_2^+)$, $C_1^+(t_1^+ \times t_2)$ and $C_2^+(t_2^+ \times t_2)$ such that

$$\{\sigma,\tau\} = \{u, Ru\} + \{0, {}^1\theta\} C_1 + \{0, {}^2\theta\} C_2,$$
$$\{\sigma^+,\tau^+\} = \{u^+,R^*u^+\} + \{0, {}^1\theta^+\} C_1^+ + \{0, {}^2\theta^+\} C_2^+.$$

We note that we may choose D and $D^+$ and hence their bases such that $C_1 = 0_{t_1}^{t_2^+}$ and $C_1^+ = 0_{t_1^+}^{t_2}$. Let us assume that we have chosen $\{\sigma,\tau\}$ and $\{\sigma^+,\tau^+\}$ in this way. From the fact that

$$\langle\{\sigma,\tau\}, \{\sigma^+,\tau^+\}\rangle = 0_{t_2}^{t_2^+}$$

it follows that

$$({}^2\theta,u^+)\, C_2 = (C_2^+)^*(u, {}^2\theta^+).$$

The matrix $(u, {}^2\theta^+)$ $(t_2^+ \times t_2^+)$ is invertible: Let $k(t_2^+ \times 1) \in \mathbb{C}^{t_2^+}$ satisfy $(u, {}^2\theta^+)k = 0_{t_2^+}^1$. Then, since

$$(u, {}^1\theta^+) = (\sigma, {}^1\theta^+) = - \langle\{\sigma,\tau\}, \{0, {}^1\theta^+\}\rangle = 0_{t_1^+}^{t_2^+},$$

it follows from (1.2) that $\{u,Ru\}\, k \in A_0$. Hence

$$\{\sigma,\tau\}\, k - \{u, Ru\}\, k = \{0, {}^2\theta\}\, C_2\, k \in A_\infty.$$

This implies that ${}^2\theta\, C_2k \in \Theta_1 \cap \Theta_2 = \{0\}$. Since ${}^2\theta$ is a basis of $\Theta_2$ we have that $C_2k = 0_{t_2}^1$ and hence

$$\{\sigma,\tau\}\, k = \{u, Ru\}\, k \in D \cap A_o = \{\{0,0\}\}.$$

Now $\{\sigma,\tau\}$ is a basis of D and therefore $k = 0^1_{t_2^+}$. This proves that $(u, {}^2\theta^+)$ is invertible. In a similar way one can prove that $({}^2\theta, u^+)(t_2 \times t_2)$ is invertible. We put

$$H = C_2(u, {}^2\theta^+)^{-1} = ({}^2\theta, u^+)^{-1}(C_2^+)^*.$$

We now prove (v). Let $\{\alpha,\beta\} \in A$. Then

$$\{\alpha,\beta\} = \{\alpha, R\alpha\} + \{0, {}^2\theta\}d + \{0, {}^1\theta\}c$$

for some $d(t_2 \times 1) \in \mathbb{C}^{t_2}$, $c(t_1 \times 1) \in \mathbb{C}^{t_1}$, and

$$(\alpha, {}^1\theta^+) = 0^1_{t_1^+}, \quad \langle\{\alpha,\beta\}, \{\sigma^+,\tau^+\}\rangle = 0^1_{t_2}.$$

The last equality implies that

$$\begin{aligned} 0^1_{t_2} &= \langle\{\alpha, R\alpha\} + \{0, {}^2\theta\}d + \{0, {}^1\theta\}c, \{u^+, R^*u^+\} + \{0, {}^2\theta^+\}C_2^+\}\rangle \\ &= -(C_2^+)^*(\alpha, {}^2\theta^+) + ({}^2\theta, u)d, \end{aligned}$$

and this shows that

$$d = H(\alpha, {}^2\theta^+).$$

Hence

$$A \subset \{\{\alpha, R\alpha + {}^2\theta H(\alpha, {}^2\theta^+) + {}^1\theta c\} \mid \alpha \in \mathcal{H}, (\alpha, {}^1\theta^+) = 0^1_{t_1^+} \text{ and } c(t_1 \times 1) \in \mathbb{C}^{t_1} \text{ is arbitrary}\}.$$

It is easy to verify that the set on the righthand side is contained in $A_1 \cap ((A^*)_\infty)^* \cap (D^+)^* = A$. Hence (v) is valid. In a similar way one can prove that $A^*$ is given by (vi).

As to the converse we note that, since R is bounded, A given by (v) is a subspace and that A clearly satisfies $A(0) = \Theta_1$. Hence $t_1 = \dim A(0)$. Since $\mathcal{D}(A_o)^\perp = A_o^*(0)$ we have that $A_o \subset A$. Clearly, $A \subset A_1$. Since also $A^*(0) = \mathcal{D}(A)^\perp = \Theta_1^+$ we see that (i) and (iv) are valid. We have shown in the beginning of this proof that (i) implies (ii). All we have left to prove is that $A^*$ equals the set on the righthand side of the equality in (vi). Let us denote that set by $A^+$. Then $A^+$ is a subspace in $\mathcal{H}^2$ satisfying $A_1^* \subset A^+ \subset A_o^*$ and it is straightforward to verify that $A^+ \subset A^*$. It is easy to see that

$$\dim(A_o^*/A^+) = t_1 + t_2^+.$$

Since $\dim(A^*/A_1^*) = \dim(A_1/A) = t_2 + t_1^+$ it follows from (1.3) that

$$\dim A_o^*/A^* = t + t^+ - (t_2 + t_1^+) = t_1 + t_2^+.$$

This shows that $A^{+} = A^{*}$, and hence that (vi) is valid.

We remark that Theorem 1.1 is a special case of Theorem 4.5 of [4]. The notation used here differs from the one used in that theorem. Here we use the letters R, $\Theta$, $\theta$, t and H instead of $\mathfrak{A}_o$, $\Phi$, $\varphi$, s and E. If in Theorem 4.5 of [4] we assume that $\mathfrak{A}_o$ is bounded and defined on all of $H$, then it follows that $d = \dim A/A_o = s_1 + s_2^{+} = t_1 + t_2^{+}$.

## 2. THE GENERALIZED INVERSE OF SUBSPACES WITH CLOSED RANGE

Let T be a subspace in $H^2$ and assume that $R(T)$ is closed. Let P, $P^{+}$ be the orthogonal projections of $H$ onto $\nu(T)$ and $\nu(T^{*})$ respectively. We define the generalized inverse $G$ of T by

$$G = (I - P)\, T^{-1}\, (I - P^{+}),$$

where I is the identity operator on $H$. In more detail, $G$ can be written as

$$G = \{\{g,f\} \in H^2 \mid \text{for some } h \in H,\ f = h - Ph,\ \{h, g - P^{+}g\} \in T\}.$$

In the following theorem we list some of the properties of $G$.

Theorem 2.1. *Let* T *be a subspace in* $H^2$ *such that* $R(T)$ *is closed, and let* $G$ *be the generalized inverse of* T. *Then:*

(i) $G$ *is a bounded linear operator defined on all of* $H$.

(ii) ("*Least square solution*"). *If* $\{h,g\} \in T$ *then* $\{Gg,g\} \in T$ *and* $\|Gg\| \leq \|h\|$.

(iii) $TG = \{\{f, f - P^{+}f\} \mid f \in H\} \dotplus T_{\infty}$, direct sum.

(iv) $GT = \{\{f, f - P f\} \mid f \in D(T)\}$.

(v) $TGT = T$.

(vi) $GTG = G$.

(vii) $G^{*}$ *is the generalized inverse of* $T^{*}$.

Proof (i) Clearly, $G$ is a linear manifold in $H^2$. If $\{0,f\} \in G$, then for some $h \in H$, $f = h - Ph$ and $\{h,0\} \in T$. It follows that $h \in \nu(T)$ and hence $f = 0$. This shows that $G$ is an operator. Let $g \in H$. Then $g - P^{+}g \in \nu(T^{*})^{\perp} = R(T)$, since $R(T)$ is closed, and so there exists an $h \in H$ such that $\{h, g - P^{+}g\} \in T$. Put $f = h - Ph$. Then $\{g,f\} \in G$, showing that $D(G) = H$. Let $(\{g_n, f_n\})$ be a sequence in $G$ converging to $\{g,f\}$ in $H^2$. In order to show that $G$ is closed we must prove that $\{g,f\} \in G$. For each $n \in \mathbb{N}$ choose $h_n \in H$ so that $f_n = h_n - Ph_n$ and $\{h_n, g_n - P^{+}g_n\} \in T$. Since $\{Ph_n, 0\} \in T$ it follows that

$$\{f_n, g_n - P^+g_n\} = \{h_n, g_n - P^+g_n\} - \{Ph_n, 0\} \in T.$$

Since T is closed and $(\{f_n, g_n - P^+g_n\})$ converges to $\{f, g - P^+g\}$ we see that $\{f, g - P^+g\} \in T$. Letting n tend to $\infty$ in $Pf_n = Ph_n - P^2h_n = 0$ we see that $Pf = 0$ and hence $f = f - Pf$. So $\{g,f\} \in G$, showing that $G$ is closed. It now follows from the Closed Graph Theorem that $G$ is bounded.

(ii) Let $\{h,g\} \in T$. Then $g \in R(T) = \nu(T^*)^{\perp}$ and hence $P^+g = 0$. It follows that $\{h, g-P^+g\} \in T$ and hence $Gg = h - Ph$. This implies that $||Gg|| \leq ||h||$ and $\{Gg, g\} = \{h,g\} - \{Ph, 0\} \in T$.

(iii) Let $\{f,g\} \in TG$. Then there exists $u \in H$ such that $\{f,u\} \in G$ and $\{u,g\} \in T$. It follows that, for some $h \in H$, $u = h - Ph$ and $\{h, f-P^+f\} \in T$. Hence

$$\{0, g - f + P^+f\} = \{u,g\} - \{h, f - P^+f\} + \{Ph, 0\} \in T,$$

which shows that $g = f - P^+f + \varphi$ for some $\varphi \in R(T_\infty)$. So $TG \subset (I - P^+) \dotplus T_\infty$. To prove the converse inclusion, let $f \in H$ and let $g = f - P^+f + \varphi$, $\varphi \in R(T_\infty)$. Since $f - P^+f \in \nu(T^*)^{\perp} = R(T)$, there is an $h \in H$ such that $\{h, f - P^+f\} \in T$. Put $u = h - Ph$. Then $\{f,u\} \in G$. Furthermore,

$$\{u,g\} = \{h, f - P^+f\} + \{0,\varphi\} - \{Ph, 0\} \in T.$$

Hence $\{f,g\} \in TG$. This shows that $TG = (I - P^+) + T_\infty$. Since $I - P^+$ is an operator and $T_\infty$ the multivalued part of T this sum is a direct sum.

(iv) - (vi). These equalities can be proved in the same way as the equality in (iii).

(vii). This follows from

$$\begin{aligned} G^* &= [(I - P)\,T^{-1}(I - P^+)]^* \\ &= (I - P^+)^*\,(T^*)^{-1}\,(I - P)^* \\ &= (I - P^+)\,(T^*)^{-1}\,(I - P). \end{aligned}$$

<u>Example</u>. Let M be a constant $n \times n$ matrix with coefficients in $\mathbb{C}$. We want to determine the generalized inverse $G$ of M; $G$ is called the <u>Moore-Penrose inverse</u> of M. We first apply Theorem 1.1. Let $H = \mathbb{C}^n$ consisting of all complex $n \times 1$ matrices and provided with the inner product $(f,g) = g^*f$. Let $A = M^{-1} = \{\{Mf, f\} \mid f \in \mathbb{C}^n\}$, and put $A_0 = \{\{0,0\}\}$ and $A_1 = H^2$. Then $A_0^* = A_1$, $A_1(0) = H$ and hence $t = t^+ = n$. Furthermore $A_1 = I_n \dotplus (A_1)_\infty$, $A_0^* = I_n \dotplus (A_0^*)_\infty$, direct sums, and $I_n^* = I_n$. This implies that we may take $R = I_n$. Let $t_1 = \dim A(0) = \dim \nu(M)$, $t_1^+ = \dim A^*(0) = \nu(M^*)$ and let ${}^1\theta(m \times t_1)$, ${}^1\theta^+ (m \times t_1^+)$ be bases for $\nu(M)$ and $\nu(M^*)$ respectively. Extend these bases to bases $({}^1\theta : {}^2\theta)$ and $({}^1\theta^+ : {}^2\theta^+)$ for $H$. Then there

exists a constant $(n - t_1) \times (n - t_1^+)$ matrix H such that

$$M^{-1} = \{\{\alpha,\ \alpha + {}^2\theta H({}^2\theta^+)^*\alpha + {}^1\theta c\} | \alpha \in H, ({}^1\theta^+)^*\alpha = O^1_{t_1^+} \text{ and } c(t_1 \times 1) \in \mathbb{C}^{t_1} \text{ is arbitrary}\}.$$

Now assume that $({}^1\theta : {}^2\theta)$ and $({}^1\theta^+ : {}^2\theta^+)$ are orthonormal. Let $\{g,f\} \in G$. Then there exist $\{u,v\} \in M^{-1}$ such that

$$u = (I - P^+)\ g = g - {}^1\theta^+({}^1\theta^+)^*g,$$

$$f = (I - P)\ \ v = v - {}^1\theta\ ({}^1\theta)^*v.$$

It follows that

$$Gg = f$$

$$= [I_n - {}^1\theta({}^1\theta^+)^* + {}^2\theta\ H({}^2\theta^+)^* - {}^1\theta({}^1\theta)^* + {}^1\theta({}^1\theta)^*\ {}^1\theta^+({}^1\theta^+)^*]g.$$

Hence

$$G = I_n - {}^1\theta({}^1\theta^+)^* + {}^2\theta H({}^2\theta^+)^* - {}^1\theta({}^1\theta)^* + {}^1\theta({}^1\theta)^*\ {}^1\theta^+({}^1\theta^+)^*.$$

We observe that ${}^2\theta^+ \in D(M^{-1})$ and hence $\{{}^2\theta^+ + {}^2\theta H,\ {}^2\theta^+\} \in M$.
Thus H satisfies

$$(2.1) \qquad (M^2\theta)\ H = (I_n - M)\ {}^2\theta^+.$$

If for some $c(t_2 \times 1) \in \mathbb{C}^{t_2}$ $M^2\theta c = O^1_n$, then ${}^2\theta\ c \in \nu(M)$, which implies that $c = O^1_{t_2}$. Hence the rank of $M^2\theta$ $(n \times t_2)$ equals $t_2$.
This proves that H is uniquely determined by (2.1).

Using exactly the same procedure as in the above example we shall construct the generalized inverse for a regular ordinary differential subspace.

## 3. REGULAR ORDINARY DIFFERENTIAL SUBSPACES IN $L^2[a,b]$

Let L be a regular ordinary differential expression of order n acting on $m \times 1$ matrix-valued functions defined on a compact real interval [a,b],

$$L = \sum_{k=o}^{n} P_k D^k, \quad D = d/dx,$$

where the $m \times m$ matrices $P_k \in C^k[a,b]$ and $P_n$ is invertible on [a,b].
The Lagrange adjoint $L^+$ of L is given by

$$L^+ = \sum_{k=o}^{n} (-1)^k D^k P_k^* = \sum_{k=o}^{n} Q_k D^k,$$

where $Q_k \in C^k[a,b]$ and $Q_n$ is invertible on $[a,b]$.

Let $H = L^2[a,b]$ be the Hilbertspace of (equivalence classes of) measurable $m \times 1$ matrix-valued functions $f : [a,b] \to ¢^m$ for which $\int_a^b f^*f < \infty$. The innerproduct is given by

$$(f,g) = \int_a^b g^*f \quad , \quad f,g \in H.$$

Let $T_o$ and $T_1$ be the <u>minimal</u> and <u>maximal operators associated with L in $H$</u>, i.e., let

$$T_1 = \{\{f,Lf\} \mid f \in C^{n-1}[a,b],\ f^{(n-1)} \in AC[a,b],\ Lf \in H\},$$

$$T_o = \{\{f,Lf\} \in T_1 \mid \tilde{f}(a) = \tilde{f}(b) = 0^1_{mn}\}.$$

Here AC $[a,b]$ is the set of all matrix-valued functions on $[a,b]$ which are absolutely continuous there, and if u is a $k \times \ell$ matrix-valued function on $[a,b]$ which is $n - 1$ times differentiable, $\tilde{u}$ is the $nk \times \ell$ matrix-valued function given by

$$\tilde{u} = \begin{pmatrix} u \\ u' \\ \vdots \\ u^{(n-1)} \end{pmatrix}.$$

The adjoints of $T_o, T_1$ are the maximal and minimal operators $T_1^+$, $T_o^+$ associated with $L^+$ in $H$,

$$T_1^+ = T_o^* = \{\{f,L^+f\} \mid f \in C^{n-1}[a,b],\ f^{(n-1)} \in AC[a,b],\ L^+f \in H\},$$

$$T_o^+ = T_1^* = \{\{f,L^+f\} \in T_1^+ \mid \tilde{f}(a) = \tilde{f}(b) = 0^1_{mn}\}.$$

It is well-known that $T_o$ and $T_1$ are closed in $H^2$, $T_o^{-1}$ and $(T_o^+)^{-1}$ are operators, $R(T_o)$ and $R(T_o^+)$ are closed in $H$ and that $R(T_1) = R(T_1^+) = H$. Green's formula states that if $\{f,g\} \in T_1$ and $\{f^+,g^+\} \in T_1^+$ then

$$\begin{aligned} \langle\{f,g\}, \{f^+,g^+\}\rangle &= (g,f^+) - (f,g^+) \\ &= [ff^+](x)\,\Big|_a^b \\ &= [ff^+](b) - [ff^+](a), \end{aligned}$$

where

$$[ff^+](x) = (\tilde{f}^+)^*(x)\, B(x)\, \tilde{f}(x)$$

and $B$ is a continuous, invertible $mn \times mn$ matrix-valued function on $[a,b]$; if $B$ is considered as an $n \times n$ matrix whose entries are given by $m \times m$ matrices, then its $\ell,j$ -th element is given by

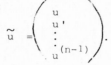

$$B_{\ell j} = \begin{cases} \sum_{k=\ell+j-1}^{n} (-1)^{k-j} \binom{k-j}{\ell-1} P_k^{(k-\ell-j+1)} & , \ \ell + j \leq n+1, \\ 0_m^m & , \ \ell + j > n+1, \end{cases}$$

and $\det B = (\det P_n)^n$.

Let $s(m \times mn)$ and $s^+ (m \times mn)$ be bases for $\nu(T_1)$ and $\nu(T_1^+)$ respectively. Let R be the integral operator defined on all of $H$ by

$$Rh(x) = \int_a^b k(x,y)\, h(y)dy \ , \ x \in [a,b], \ h \in H,$$

with kernel

$$k(x,y) = \begin{cases} s(x) \, [ss^+]^{-1} (s^+)^* (y) & , \ a \leq y \leq x \leq b, \\ 0_m^m & , \ a \leq x < y \leq b. \end{cases}$$

Here $[ss^+] = [ss^+]\,(x) = (\tilde{s}^+)^*(x)\, B(x)\, \tilde{s}(x)$ is independent of $x \in [a,b]$ and invertible. Clearly R is a bounded operator. It is a rightinverse of $T_1$, i.e., $TRh = h$ for all $h \in H$ and

$$T_o^{-1} = \{\{f, Rf\} \mid f \in R(T_o)\}.$$

The adjoint $R^*$ of R is the bounded integral operator defined on $H$ by

$$R^*h(x) = \int_a^b k^+(x,y)\, h(y)dy, \ x \in [a,b], \ h \in H,$$

with kernel

$$k^+(x,y) = k^*(y,x) \ , \ x \ y \in [a,b].$$

$R^*$ is a rightinverse of $T_1^+$ and

$$(T_o^+)^{-1} = \{\{f, R^*f\} \mid f \in R(T_o^+)\}.$$

Let B, $B^+$ be finite dimensional subspaces in $H^2$. A <u>regular ordinary differential subspace</u> T in $H^2$ <u>associated with</u> L, B <u>and</u> $B^+$ is a subspace T in $H^2$ satisfying

$$(3.1) \qquad T_o \cap (B^+)^* \subset T \subset T_1 \dot{+} B.$$

We may and shall always choose B and $B^+$ so that $T_1 \dot{+} B$ and $T_1^+ + B^+$ are direct sums. Note that if T satisfies (3.1) then $T^*$ is a regular ordinary differential subspace in $H^2$ associated with $L^+$, $B^+$ and B, for (3.1) implies that

$$T_o^+ \cap B^* \subset T^* \subset T_1^+ \dotplus B^+.$$

Let T satisfy (3.1) and put

$$p = \dim B \quad , \quad p^+ = \dim B^+.$$

Since $\dim (T_1/T_o) = 2mn$, it follows that

$$\dim(T_1 \dotplus B)/(T_o \cap (B^+)^*) = \dim(T_1^+ \dotplus B^+)/(T_o^+ \cap B^*) = 2mn + p + p^+.$$

This implies that there exist finite dimensional subspaces C, $C^+$ in $H^2$ such that

(3.2) $\quad T = (T_o \cap (B^+)^*) \dotplus C \quad , \quad T^* = (T_o^+ \cap B^*) \dotplus C^+$, direct sums,

or, equivalently, such that

(3.3) $\quad T = (T_1 \dotplus B) \cap (C^+)^* \; , \; T^* = (T_1^+ \dotplus B^+) \cap C^*.$

The equalities in (3.3) show that T, $T^*$ may be considered as being restrictions of $T \dotplus B$ and $T^+ \dotplus B^+$ respectively by means of the subspaces $C^+$ and C. Indeed, the subspaces $C^+$ and C represent the side conditions of T and $T^*$ which are a mixture of two point boundary and integral conditions. In [4; Theorem 9.2] these side conditions are given explicitly. We refer to Section 5 for some examples.

The first equality in (3.2) implies that

$$R(T) = R(T_o \cap (B^+)^*) + R(C).$$

Since $R(T_1^+ \dotplus B^+) = R(T_1^+) = H$ it follows that $R(T_o \cap (B^+)^*)$ is closed. Since $\dim R(C) < \infty$ it follows that $R(T)$ is closed. Hence T possesses a generalized inverse, and so does $T^*$.

## 4. THE GENERALIZED INVERSE FOR REGULAR ORDINARY DIFFERENTIAL SUBSPACES

In this section we consider a fixed regular ordinary differential subspace T in $H^2$ associated with L, B and $B^+$. Thus we have the following inclusions

$$(4.1) \quad \begin{cases} T_o \cap (B^+)^* \subset T \subset T_1 \dotplus B, \\ T_o^+ \cap B^* \subset T^* \subset T_1^+ \dotplus B^+, \end{cases}$$

where B, $B^+$ are fixed finite dimensional subspaces in $H^2$ such that

(4.2) $\quad T_1 \dotplus B \; , \; T_1^+ \dotplus B^+$ are direct sums.

We put

(4.3) $\quad p = \dim B \; , \; p^+ = \dim B^+.$

Let $G$ denote the generalized inverse of T. In order to characterize $G$ and $G^*$ we are going to determine suitable representations for $T^{-1}$ and $(T^*)^{-1}$. This we do by applying Theorem 1.1.

Let $\{\sigma,\tau\}$ $(m \times p)$, $\{\sigma^+,\tau^+\}$ $(m \times p^+)$ be bases for B and $B^+$ respectively. We assume that $\sigma$, $\tau$, $\sigma^+$ and $\tau^+$ are defined at every $x \in [a,b]$. Let

$$\rho = \sigma - R\tau \ (m \times p) \quad , \quad \rho^+ = \sigma^+ - R^*\tau^+ \ (m \times p^+) ,$$

where R, $R^*$ are the integral operators associated with L, $L^+$ defined in Section 3. Then

$$(4.4) \quad \begin{cases} (s : \rho) \ (m \times (mn + p)) & \text{is a basis for } \nu(T_1 \dot{+} B), \\ (s^+ : \rho^+) (m \times (mn + p^+)) & \text{is a basis for } \nu(T_1^+ \dot{+} B^+). \end{cases}$$

We shall prove only the first part of (4.4). The second part can be proved in the same way. Let $f \in \nu\ (T \dot{+} B)$. Then $\{f,0\} \in T_1 \dot{+} B$ and hence there exist $\{u,v\} \in T_1$ and $c(p \times 1) \in ¢^p$ such that $\{f,0\} = \{u,v\} + \{\sigma,\tau\}c$. It follows that $T_1 u = v = -\tau c = -T_1 R\ \tau\ c$. Hence there exists a $d(mn \times 1) \in ¢^{mn}$ such that $u + R\,\tau\, c = sd$. This implies that

$$f = u + \sigma c = (s : \rho)\ \binom{d}{c} \in \text{span } (s : \rho).$$

Conversely, if $c \in ¢^p$, $d \in ¢^{mn}$, then

$$(4.5) \quad \{sd + \rho c, 0\} = \{s,0\}d - \{R\tau,\tau\}c + \{\sigma,\tau\}c \in T_1 \dot{+} B$$

and hence

$$sd + \rho c \in \nu(T_1 \dot{+} B).$$

This proves that $\nu(T_1 \dot{+} B) = \text{span } (s : \rho)$. Suppose that for some $c \in ¢^p$, $d \in ¢^{mn}$, $sd + \rho c = 0_m^1$. Then it follows from (4.5) that

$$\{\sigma,\tau\}c = \{R\tau,\tau\}c - \{s,0\}d \in T_1 \cap B = \{\{0,0\}\}.$$

Since $\{\sigma,\tau\}$ is a basis for B, $c = 0_p^1$ and hence $\{s,0\}d = \{0,0\}$. Since s is a basis for $\nu(T_1)$ we have that $d = 0_{mn}^1$. This proves that $(s : \rho)$ is linearly independent.

<u>Theorem 4.1.</u> <u>Let</u> T <u>be a regular ordinary differential subspace in</u> $H^2$ <u>associated with</u> L,B <u>and</u> $B^+$ <u>described by</u> (4.1), (4.2) <u>and</u> (4.3), <u>with</u> $t_1 = \dim \nu(T)$, and $t_1^+ = \dim \nu(T^*)$. <u>Then</u>

(i) $\dim(T/(T_0 \cap (B^+)^*)) = t_1 + t_2^+$, $\dim(T^*/(T_0^+ \cap B^*)) = t_2 + t_1^+$ <u>where</u> $t_1 + t_2 = mn + p$, $t_1^+ + t_2^+ = mn + p^+$. <u>Let</u>

$$\text{(ii)} \begin{cases} \nu(T_1 \dotplus B) = \Theta_1 + \Theta_2 , & \nu(T_1^{+} \dotplus B^{+}) = \Theta_1^{+} + \Theta_2^{+}, \text{ direct sums,} \\ \dim \Theta_j = t_j , & \dim \Theta_j^{+} = t_j^{+} , \quad j = 1,2, \end{cases}$$

be decompositions of $\nu(T_1 \dotplus B)$, $\nu(T_1^{+} \dotplus B^{+})$ into linear manifolds in $H$ such that

(iii) $\nu(T) = \Theta_1$ , $\nu(T^{*}) = \Theta_1^{+}$.

Let ${}^{j}\theta$, ${}^{j}\theta^{+}$ be bases for $\Theta_j$, $\Theta_j^{+}$, $j = 1,2$, respectively. Then there exists a constant $t_2 \times t_2^{+}$ matrix H such that

(iv) $T^{-1} = \{\{\alpha, R\alpha + {}^{2}\theta H (\alpha, {}^{2}\theta^{+}) + {}^{1}\theta c\} \mid \alpha \in H,\ (\alpha, {}^{1}\theta^{+}) = 0^{1}_{t_1}$ and $c(t_1 \times 1) \in \mathbb{C}^{t_1}$ is arbitrary$\}$,

and

(v) $(T^{*})^{-1} = \{\{\alpha^{+}, R^{*}\alpha^{+} + {}^{2}\theta^{+} H^{*}(\alpha_1^{+}, {}^{2}\theta) + {}^{1}\theta^{+}c^{+}\} \mid \alpha^{+} \in H, (\alpha^{+}, {}^{1}\theta) = 0^{1}_{t_1}$ and $c(t_1^{+} \times 1) \in \mathbb{C}^{t_1^{+}}$ is arbitrary$\}$,

where R, $R^{*}$ are the integral operators associated with L,$L^{+}$ defined in Section 3.

Conversely, if (ii) represent decompositions of $\nu(T_1 \dotplus B)$, $\nu(T_1^{+} \dotplus B^{+})$ and ${}^{j}\theta$, ${}^{j}\theta^{+}$ are bases for $\Theta_j$, $\Theta_j^{+}$, $j = 1,2$, respectively and $T^{-1}$ is defined by (iv), where H is any constant $t_2 \times t_2^{+}$ matrix, then T is a regular ordinary differential subspace in $H^2$ associated with L, B and $B^{+}$ satisfying (i), (iii) and (v).

Proof. Put $A_o = (T_o \cap (B^{+})^{*})^{-1}$ and $A_1 = (T_1 \dotplus B)^{-1}$. Then $A_o \subset A_1$ and, according to (4.4),

$$t = \dim A_1(0) = \dim \nu(T_1 \dotplus B) = mn + p,$$
$$t^{+} = \dim A_o^{*}(0) = \dim \nu(T_1^{+} \dotplus B^{+}) = mn + p^{+}.$$

Furthermore,

$$A_1 = R \dotplus (A_1)_\infty , \quad A_o^{*} = R^{*} \dotplus (A_o^{*})_\infty, \text{ direct sums.}$$

We will prove the first equality. Since $R \subset T_1$ and $A_1(0) = \nu(T \dotplus B)$ we have that $R \dotplus (A_1)_\infty \subset A_1$. To prove the converse inclusion, let $\{f,g\} \in A_1$. Then $\{g,f\} \in T_1 \dotplus B$ and there exist elements $\{u,v\} \in T_1$ and c $(p \times 1) \in \mathbb{C}^{p}$ such that

$$\{g,f\} = \{u,v\} + \{\sigma,\tau\}\, c.$$

Since $T_1 u = v = T_1 Rv$, $u = Rv + sd$ for some d $(mn \times 1) \in \mathbb{C}^{mn}$.

Hence

$$\{f,g\} = \{v,Rv\} - \{\tau,R\tau\}\, c + \{0,sd + \rho c\} \in R \dotplus (A_1)_\infty.$$

The second equality can be proved in the same way. That the sums here are direct follows from the fact that R, $R^*$ are operators.

Putting $A = T^{-1}$ we see that the above theorem follows from Theorem 1.1.

Remark. The matrix H appearing in the above theorem can be determined from the boundary conditions which make T a restriction of $T_1 \dotplus B$. To see this, let $C^+$ be a finite dimensional subspace in $H^2$ such that

$$T = (T_1 \dotplus B) \cap (C^+)^*, T^* = (T_o^+ \cap B^*) \dotplus C^+, \text{ direct sum,}$$

(See (3.2) and (3.3)). Then $\dim C^+ = t_2 + t_1^+$. Let $\{\mu^+,\nu^+\}(m \times (t_2 + t_1^+))$ be a basis for $C^+$. Choose the decompositions of $\nu(T_1^+ \dotplus B^+)$ and the bases ${}^j\theta^+$, j = 1,2 such that

$$({}^2\theta^+, {}^2\theta^+) = I_{t_2^+} \quad , \quad ({}^2\theta^+ , {}^1\theta^+) = 0_{t_1^+}^{t_2^+} .$$

Then it follows that ${}^2\theta^+ \; (m \times t_2^+) \in D(T^{-1})$ which implies that

$$\{R^2\theta^+ + {}^2\theta H, {}^2\theta^+\} \in T.$$

Hence

$$<\{R^2\theta^+ + {}^2\theta H, {}^2\theta^+\} , \{\mu^+,\nu^+\}> = 0_{t_2 + t_1^+}^{t_2^+} ,$$

and this implies that

$$({}^2\theta,\nu^+)H = <\{R^2\,\theta^+, {}^2\theta^+\} , \{\mu^+,\nu^+\}>.$$

We claim that

$$\text{rank} \quad ({}^2\theta, \nu^+) = t_2.$$

Since $({}^2\theta, \nu^+)$ is a $(t_2 + t_1^+) \times t_2$ matrix we have that the rank of this matrix does not exceed $t_2$. Suppose that, for some c $(t_2 \times 1) \in \mathbb{C}^{t_2}$, $({}^2\theta, \nu^+)\, \mathbf{c} = 0_{t_2 + t_1^+}^1$. Then

$$<\{{}^2\theta c, 0\} , \{\mu^+,\nu^+\} > = 0_{t_2 + t_1^+}^1 ,$$

showing that ${}^2\theta\, c \in \Theta_1 \cap \Theta_2 = \{0\}$. Hence $c = 0_{t_2}^1$. This proves the claim. It follows that there exists a constant $(t_2 + t_1^+) \times (t_2 + t_1^+)$ matrix $H_o$ such that

$$H_o \, ({}^2\theta, \nu^+) = \begin{pmatrix} 0_{t_1^+}^{t_2} \\ I_{t_2} \end{pmatrix}.$$

Thus H is determined by

$$H = (O_{t_2}^{t_1^+} : I_{t_2}) \, H_o <\{R^2\theta^+, {}^2\theta^+\} , \{\mu^+,\nu^+\}>.$$

The following theorem characterizes the generalized inverse of the subspace T.

Theorem 4.2. Let T be a regular differential subspace in $H^2$ described in the first part of Theorem 4.1. Assume that the bases $({}^1\theta : {}^2\theta)$ and $({}^1\theta^+ : {}^2\theta^+)$ are orthonormal. Then the generalized inverse $G$ of T is an integral operator,

$$Gg(x) = \int_a G(x,y) \, g(y)dy, \text{ a.a } x \in [a,b],$$

for all $g \in H$ with kernel

$$\text{(i) } G(x,y) = k(x,y) - (R^1\theta^+)(x)({}^1\theta^+)^*(y) + {}^2\theta(x) \, H({}^2\theta^+)^*(y)$$
$$- {}^1\theta(x)(R^* \, {}^1\theta)^*(y) + {}^1\theta(x)(R^1\theta^+ , {}^1\theta)({}^1\theta^+)^*(y), \; x,y \in [a,b] ,$$

where k is the kernel of the integral operator R and H is the $t_2 \times t_2^+$ matrix of Theorem 4.1. Furthermore $G^*$ is an integral operator with kernel

$$\text{(ii) } G^+(x,y) = G^*(y,x), \quad x, y \in [a,b].$$

Proof. Let P, $P^+$ be the orthogonal projections from $H$ onto $\nu(T)$ and $\nu(T^*)$ respectively. Then $G = (I - P) \, T^{-1} \, (I - P^+)$. Let $\{g,f\} \in G$. Then there exist $u, v \in H$ such that

$$u = (I - P^+)g, \quad \{u,v\} \in T^{-1} \quad , \quad f = (I - P)v.$$

According to Theorem 4.1 there exist $\alpha \in H$ and $c(t_1 \times 1) \in \mathbb{C}^{t_1}$ with $(\alpha, {}^1\theta^+) = O_{t_1^+}^1$ such that

$$\{u,v\} = \{\alpha, R\alpha + {}^2\theta H(\alpha, {}^2\theta^+) + {}^1\theta c\} .$$

Hence, using the orthonormality of the bases $({}^1\theta : {}^2\theta)$ and $({}^1\theta^+ : {}^2\theta^+)$, we see that

$$\alpha = u = g - {}^1\theta^+(g, {}^1\theta^+),$$
$$\begin{aligned} f &= (I - P)v \\ &= (I - P) \, (R\alpha + {}^2\theta H(\alpha, {}^2\theta^+) + {}^1\theta c) \\ &= R\alpha + {}^2\theta H \, (\alpha, {}^2\theta^+) + {}^1\theta c - {}^1\theta(R\alpha, {}^1\theta) - {}^1\theta({}^1\theta, {}^1\theta)c \\ &= R\alpha + {}^2\theta H \, (\alpha, {}^2\theta^+) - {}^1\theta(R\alpha, {}^1\theta), \end{aligned}$$

and hence that

$$f = Rg - R^1\theta(g,{}^1\theta^+) + {}^2\theta H(g,{}^2\theta^+) - {}^1\theta(Rg,{}^1\theta)$$
$$+ {}^1\theta(R^1\theta^+,\ {}^1\theta)\ (g,\ {}^1\theta^+)$$
$$= \int G(.,y)\ g(y)\ dy,$$

where G is given by (i). Clearly $G \in L^2([a,b] \times [a,b]$ and hence the last part of the theorem follows from Fubini's Theorem on interchanging the order of integration in double integrals.

The kernel G of the generalized inverse $G$ of T described in Theorem 4.2 is called the generalized Green's matrix-valued function associated with T. Concerning the uniqueness of G we have the following result.

Theorem 4.3. Let T be a regular differential subspace in $H^2$ as in Theorem 4.1. Assume that the bases $({}^1\theta : {}^2\theta)$ and $({}^1\theta^+ : {}^2\theta^+)$ are orthonormal. Let G be as in (i) of Theorem 4.2. Then the m × m matrix valued function $G_o \in L^2([a,b] \times [a,b])$ satisfies

$$G_o(x,y) = G(x,y) \quad \text{a.e. on } [a,b] \times [a,b],$$

if and only if for a.a. $y \in [a,b]$.

$$\text{(i)}\ \{G_o(.,y),\ -{}^1\theta^+({}^1\theta^+)^*(y)\} - \{k(.,y),0\} +$$
$$+ \{R({}^1\theta^+ : {}^2\theta^+),\ ({}^1\theta^+ : {}^2\theta^+)\}\ ({}^1\theta^+ : {}^2\theta^+)^*(y) \in T,$$

and

$$\text{(ii)}\ (G_o(.,y),\ {}^1\theta) = 0^m_{t_1}.$$

Proof By Theorem 4.2 (i) and Theorem 4.1 we have that for almost all $y \in [a,b]$,

$$\{G(.,y),\ -{}^1\theta^+({}^1\theta^+)^*(y)\} - \{k(.,y),0\}$$
$$+ \{R({}^1\theta^+ : {}^2\theta^+),({}^1\theta^+ : {}^2\theta^+)\}\ ({}^1\theta^+ : {}^2\theta^+)^*(y)$$
$$= \{(R\ {}^2\theta^+)({}^2\theta^+)^*(y) + {}^2\theta H({}^2\theta^+)^*(y) + {}^1\theta\ c,\ {}^2\theta^+({}^2\theta^+)^*(y)\} \in T$$

where

$$c = -(R^{*1}\theta)^*(y) + (R^1\theta^+,\ {}^1\theta)({}^1\theta^+)^*(y)$$

This shows that G satisfies (i). From $G^*\ {}^1\theta = 0^{t_1}_m$ it follows that G satisfies (ii). Hence if $G_o \in L^2([a,b] \times [a,b])$ and $G_o = G$ a.e., then $G_o$ satisfies (i) and (ii) for almost all $y \in [a,b]$. Conversely, let $G_o \in L^2([a,b] \times [a,b])$ and suppose that $G_o$ satisfies (i) and (ii) for almost all $y \in [a,b]$. Fix $y \in [a,b]$ so that (i) and (ii) are valid. Then

$$\{G_o(.,y) - G(.,y), 0\} \in T,$$

which implies that

$$G_o(.,y) - G(.,y) = {}^1\theta c$$

for some constant matrix c ($t_1 \times m$). By (ii) $c = O_{t_1}^m$. Hence

$$G_o(.,y) = G(.,y) \text{ a.e. on } [a,b] .$$

It follows that

$$\int_a^b (G_o(x,y) - G(x,y))^* (G_o(x,y) - G(x,y)dx = O_m^m .$$

This now is valid for almost all $y \in [a,b]$. Hence

$$\int_a^b \int_a^b (G_o(x,y) - G(x,y))^* (G_o(x,y) - G(x,y)dxdy = O_m^m ,$$

which shows that $G_o = G$ a.e on $[a,b] \times [a,b]$.

Remarks .1. We note that Theorems 4.1, 4.2 and 4.3 remain valid if we replace R and its kernel k defined in Section 3 by any integral operator R with kernel k which is a right inverse for $T_1$.

2. If in Theorem 4.3 (i) we replace the term

$$\{R({}^1\theta^+ : {}^2\theta^+), ({}^1\theta^+ : {}^2\theta^+)\} ({}^1\theta^+ : {}^2\theta^+)^*(y)$$

by

$$\{R\,\theta^+, \theta^+\} (\theta^+)^* (y),$$

where $\theta^+$ ($m \times t^+$) is an orthonormal basis for $\nu(T_1^+ \dot{+} B^+)$, then Theorem 4.3 remains valid.

## 5. EXAMPLES

In this section we shall without further comment make use of the notations used in the previous sections.

Example 5.1. (Two point boundary conditions). Take $B = B^+ = \{\{0,0\}\}$. Let T be a subspace satisfying $T_o \subset T \subset T_1$ and $d = \dim(T/T_o)$. Then $d = t_1 + t_2^+$ and there exist constant matrices $M^+((2mn - d) \times mn)$, $N^+$ $((2mn - d) \times mn)$, M ($d \times mn$) and N = ($d \times mn$) such that

$$\operatorname{rank}\,(M^+ : N^+) = 2\,mn - d,\ \operatorname{rank}\,(M : N) = d,$$
$$M^+ \mathcal{B}^{-1}(a)M^* - N^+ \mathcal{B}^{-1}(b)N^* = O^d_{2mn-d}\,,$$
$$T = \{\{f,Lf\} \in T_1 \mid M^+ \tilde{f}(a) + N^+ \tilde{f}(b) = O^1_{2mn-d}\},$$

and

$$T^* = \{\{f,L^+f\} \in T_1{}^+ \mid M\,\tilde{f}(a) + N\,\tilde{f}(b) = O^1_d\}.$$

See [4; Theorem 9.2]. According to the Remark after the proof of Theorem 4.1 the matrix H can be determined from the equality

$$(M^+ (\widetilde{{}^2\theta})(a) + N^+(\widetilde{{}^2\theta})(b))H = -N^+\ (\widetilde{R\,{}^2\theta^+})\ (b)$$
$$= -N^+ \mathcal{B}^{-1}(b)(\widetilde{({}^1\theta^+ : {}^2\theta^+)^*})^{-1}(b)\begin{pmatrix} & t_2 \\ O_{t_1^+} & \\ & \\ I_{t_2^+} & \end{pmatrix}^+.$$

(The matrix $H_o$ also appears in Locker's construction of the generalized inverse for differential operators with two point boundary conditions; see ([5] Lemma 2)). In [6] Locker proves various properties of the Green's function. All his results can easily be obtained from Theorem 4.3 (i). For specific examples of Green's functions for this case we refer to [6] and [8].

Example 5.2. (Stieltjes boundary conditions). In [2] Brown considers the case where $L = D + P_o$ on $[a,b]$ and

$$T = \{\{f,Lf\} \in T_1 \mid \int_a^b d\mu^*\ f = O^1_k\}\ ,$$

where $\mu$ is an $m \times k$ matrix-valued function of bounded variation on $[a,b]$. Clearly, $T \subset T_1$ and hence we may take $B = \{\{0,0\}\}$. In [4; Theorem 9.3 and its proof] it is shown that if $B^+ = \operatorname{span}\{\mu - R^*P_o{}^*\mu, 0\}$ then $T_o \cap (B^+)^* \subset T \subset T_1$. Hence $T_o{}^+ \subset T^* \subset T_1{}^+ \dot{+} B^+$. From Theorem 4.3 (i) one can now easily deduce the properties of the Green's function for this operator which are described in [2] for the Hilbertspace case.

We now give some specific examples

Example 5.3. Let $n = m = 1$ and $L = D$ on $[0,1]$. Consider

$$T = \{\{f,f'\} \in T_1 \mid f(0) + f(\tfrac{1}{2}) + f(1) = 0\}.$$

Let

$$\sigma^+(x) = \begin{cases} 4x-1 , & x \in [0,\frac{1}{2}] \\ 2x-1 & x \in (\frac{1}{2},1] \end{cases} \qquad \tau^+(x) = \begin{cases} -4 , & x \in [0,\frac{1}{2}] \\ -2 , & x \in (\frac{1}{2},1] \end{cases}$$

and $B^+ = \text{span}\,\{\sigma^+,\tau^+\}$. Then $T_o \cap (B^+)^* \subset T = T_1 \cap (B^+)^* \subset T_1$ and hence $T_o^+ \subset T^* = T_o^+ \dot{+} B^+ \subset T_1^+ \dot{+} B^+$, direct sums. It is now easy to verify that for all $x \in [0,1]$

$${}^1\theta(x) = 0, \qquad {}^2\theta(x) = 1,$$

$${}^1\theta^+(x) = 0, \qquad {}^2\theta^+(x) = \begin{cases} (1,\ 1) & x \in [0,\frac{1}{2}] , \\ (1,-1) & x \in (\frac{1}{2},1] . \end{cases}$$

H can be determined from

$$({}^2\theta(0) + {}^2\theta(\tfrac{1}{2}) + {}^2\theta(1))H = -\ (R\ {}^2\theta^+(0) + R\ {}^2\theta^+(\tfrac{1}{2}) + R\ {}^2\theta^+(1)).$$

We find $H = -(\frac{1}{2},\frac{1}{6})$. Since

$$k(x,y) = \begin{cases} 1 , & 0 \leqq y \leqq x \leqq 1, \\ 0 , & 0 \leqq x < y \leqq 1, \end{cases}$$

it follows from Theorem 4.2 (i) that

$$G(x,y) = \begin{cases} 1/3 & x \in [0,\frac{1}{2}] \quad y \in [0,\frac{1}{2}] , \ y \leqq x, \\ -2/3 & x \in [0,\frac{1}{2}] \quad y \in [0,\frac{1}{2}] , \ x < y, \\ 2/3 & x \in (\frac{1}{2},1] \quad y \in (\frac{1}{2},1] , \ y \leqq x, \\ -1/3 & x \in (\frac{1}{2},1] \quad y \in (\frac{1}{2},1] , \ x < y, \\ -1/3 & x \in [0,\frac{1}{2}] \quad y \in (\frac{1}{2},1] , \\ 1/3 & x \in (\frac{1}{2},1] \quad y \in [0,\frac{1}{2}] . \end{cases}$$

Note that $\frac{\partial}{\partial x} G(x,y) = 0$, $x \neq y$, $G(0,y) + G(\frac{1}{2},y) + G(1,y) = 0$ for all $y \in [a,b]$ and that $G(y^+,y) - G(y^-,y) = 1$ for all $y \in [a,b]$, provided we take $y^- = a$ if $y = a$ and $y^+ = b$ if $y = b$. Also, $G^+(x,y) = G\ (y,x)$ is the generalized Green's function for

$$\begin{aligned} T^* &= T_o^+ + B^+ \\ &= \{\{f,g\} \mid f \in AC[0,\tfrac{1}{2}],\ f \in AC[\tfrac{1}{2},1],\ g = -f' \text{ on } [0,\tfrac{1}{2}] \text{ and} \\ &\qquad \text{on } [\tfrac{1}{2},1],\ g \in H \text{ and } f(0) + f(1) = 0\}. \end{aligned}$$

Example 5.4. Let $n = m = 1$ and $L = D$ on $[0,1]$. Consider

$$T = \{\{f, f' + g\int_0^1 f\} \mid \{f,f'\} \in T_1\}, \quad g(x) = x \text{ on } [0,1].$$

Let

$$\sigma(x) = 0, \quad \tau(x) = x, \quad x \in [0,1],$$
$$\sigma^+(x) = 0, \quad \tau^+(x) = 1, \quad x \in [0,1],$$
$$\mu^+(x) = 12x(1-x), \quad \nu^+(x) = 24x-12, \quad x \in [0,1],$$

and $B = \text{span}\,\{\sigma,\tau\}$, $B^+ = \text{span}\,\{\sigma^+,\tau^+\}$ and $C^+ = \text{span}\,\{\mu^+,\nu^+\}$. Then $T_0 \cap (B^+)^* \subset T = (T_1 \dot{+} B) \cap (C^+)^* \subset T_1 \dot{+} B$ and hence $T_0^+ \cap B^* \subset T^* = (T_0^+ \cap B^*) \dot{+} C^+ \subset T_1^+ \dot{+} B^+$. From this we find that for $x \in [0,1]$

$$^1\theta(x) = \sqrt{\frac{5}{184}}\,(3x^2 - 7), \qquad ^2\theta(x) = \frac{1}{\sqrt{184}}\,(45x^2 - 13),$$

$$^1\theta^+(x) = 0, \qquad ^2\theta^+(x) = (1, \sqrt{3}\,(2x-1)),$$

and that

$$H = \frac{1}{\sqrt{184}}\,(-1, \frac{1}{\sqrt{3}}).$$

Hence by Theorem 4.2 (i)

$$G(x,y) = \begin{cases} 1 + \dfrac{1}{184}\,[2(45x^2-13)(y-1) + 5(3x^2-7)(y^3-7y+6)], & y \leq x, \\ \dfrac{1}{184}\,[2(45x^2-13)(y-1) + 5(3x^2-7)(y^3-7y+6)], & x < y. \end{cases}$$

By Theorem 4.3 (i) we have that

$$\{G(.,y),0\} - \{k(.,y),0\} + \{h(.,y), h'(.,y)\} \in T, \quad y \in [0,1],$$

where

$$h(x,y) = 6x^2y - 6xy - 3x^2 + 4x, \quad h'(x,y) = \frac{\partial}{\partial x} h(x,y), \quad x,y \in [0,1].$$

From this it easily follows that

$$\frac{\partial}{\partial x} G(x,y) + x\int_0^1 G(t,y)\,dt = 0, \quad x \neq y.$$

Note that $G^+(x,y) = G(y,x)$ is the generalized Green's function for

$$T^* = \{\{u, -u' + \int_0^1 gu\} \mid \{u,-u'\} \in T_0^+\}, \quad g(x) = x, \ x \in [0,1],$$

and that $G^* = (T^*)^{-1}$.

Concerning Green's functions and differential subspaces with general boundary conditions we refer to the survey papers of Conti [11] and Krall [12]. For more information about generalized inverses we refer to [13].

References.

1. J.S. Bradley, Generalized Green's matrices for compatible differential systems, Michigan Math.J. 13 (1966) 97-108.
2. R.C. Brown, Generalized Green's functions and generalized inverses for linear differential systems with Stieltjes boundary conditions, J.Diff.Equations 16(1974) 335-351.
3. H. Chitwood, Generalized Green's matrices for linear differential systems, Siam. J. Math. Anal. 4 (1973) 104-110.
4. E.A. Coddington and A. Dijksma, Adjoint subspaces in Banach spaces, with applications to ordinary differential subspaces, to appear in Ann. di. Mat. Pura ed Appl.
5. J. Locker, An existence analysis for nonlinear boundary-value problems, Siam.J. Appl. Math. 19 (1970) 199-207.
6. J. Locker, The generalized Green's function for an n-th order linear differential operator, Transactions Amer. Math. Soc. 228 (1977) 243-268.
7. W.S. Loud, Generalized inverses and generalized Green's functions. Siam. J. Appl. Math. 14 (1966) 342-369.
8. W.S. Loud, Some examples of generalized Green's functions and generalized Green's matrices, Siam. Review 12 (1970) 194-210.
9. W.T. Reid, Generalized Green's matrices for compatible systems of differential equations, Amer. J. Math. 53(1931) 443-459.
10. W.T. Reid, Generalized Green's matrices for two-point boundary value problems, Siam. J. Appl. Math. 15 (1967) 856-870.
11. R. Conti, Recent trends in the theory of boundary value problems for ordinary differential equations, Boll. U.M.I. (3) XXII (1967), 135-178.
12. A.M. Krall, The development of general differential and general differential-boundary systems, Rocky Mountain J. Math. 5 (1975) 493-542.
13. M.Z. Nashed (Editor), Generalized Inverses and Applications, Academic Press, New York 1976.

Department of Mathematics
Rijksuniversiteit Groningen
Groningen, The Netherlands

*Differential Equations and Applications*
*W. Eckhaus and E.M. de Jager (eds.)*


# A GENERALIZATION OF HARTOGS THEOREM

J. BESJES and R. MARTINI

## §1. INTRODUCTION

In the theory of complex functions of several complex variables there is a basic result of Hartogs that a separately analytic function is analytic. From the point of view of linear partial differential equations this may be stated as follows.

A separately continuous differentiable function u, say, of two independent complex variables $z_1$ and $z_2$ which satisfies the Cauchy-Riemann equations

$$\frac{\partial u}{\partial \bar{z}_1} = 0, \quad \frac{\partial u}{\partial \bar{z}_2} = 0$$

is analytic in both variables $z_1$ and $z_2$ together.

More generally we may ask whether a separately regular function u of two independent space-variables x and y which satisfies the equations

$$L_x u = 0, \quad L_y u = 0 ,$$

where $L_x$ respectively $L_y$ are elliptic differential operators in x-space and y-space, is automatically regular in the variables x and y together.

In case $L_x$ equals the Cauchy-Riemann operator and $L_y$ the Laplacian or (more generally) the iterated Laplacian then the answer is affirmative and this paper contains a proof of this.

## §2. SOME CLASSICAL RESULTS

In the following sections we shall need the following variant of a famous theorem on sequences of analytic functions of one complex variable, due to Hartogs (1906).

Theorem 1. Let $(a_\nu)$, $\nu = 1,2,3,\ldots$, be a sequence of functions of one complex variable, defined and analytic on the disc $\Omega = \{z \mid |z| \le R\}$.
Suppose that there exist constants $M, r_0, r_1$, $0 < r_1 < r_0$ such that

$$\lim_{\nu\to\infty} a_\nu(z) r_0^\nu = 0 \quad \text{for any fixed} \quad z \in \Omega$$

and

$$|a_\nu(z) r_1^\nu| \le M \quad \text{for all} \quad z \in \Omega \quad \text{and all} \quad \nu = 1,2,\ldots .$$

Then for any compact subset $K \subset \Omega$ and for any $r$ with $0 < r < r_0$ there exists an index $\nu(K,r)$ such that

$$|a_\nu(z) r^\nu| \le M \quad \text{for all} \quad z \in K \quad \text{and all} \quad \nu \ge \nu(K,r).$$

Proof. It is no restriction to assume that $r_1 = 1$ and

$$K = \{z \mid |z| \le R - 3\delta\} \quad \text{with} \quad 0 < \delta < \tfrac{1}{3}R.$$

We consider functions $(b_\nu)$, $\nu = 1,2,3,\ldots$, defined by

$$b_\nu(z) = \frac{1}{\nu} \log \frac{|a_\nu(z)|}{M}$$

$(\log 0 = -\infty)$. Now $\overline{\lim_{\nu\to\infty}} |a_\nu(z) r_0^\nu|^{1/\nu} \le 1$ and the logarithm is a strictly increasing function so it follows that
$\overline{\lim_{\nu\to\infty}} b_\nu(z) \le -\log r_0$ for any $z \in \Omega$.
The assumption $r_1 = 1$ implies that

$$b_\nu(z) \le 0 \quad \text{for any} \quad z \in \Omega.$$

Let $z \in \Omega$ be arbitrary but fixed. Then by Fatou's lemma (see e.g. RUDIN

[3], p. 22) we have

$$\varlimsup_{\nu\to\infty} \int_{|z-z'|<\delta} b_\nu(x',y')dx'dy' \le \int_{|z-z'|<\delta} \varlimsup_{\nu\to\infty} b_\nu(x',y')dx'dy' \le$$

$$\le \int_{|z-z'|<\delta} - \log r_0 dx'dy' = -\pi\delta^2 \log r_0$$

with the notation: $(x',y') = z'$, $b_\nu(x',y') = b_\nu(z')$. From this inequality it follows that for any $\varepsilon > 0$ there exists an index $\nu_0(\varepsilon,\delta)$ such that

$$\int_{|z-z'|<\delta} b_\nu(x',y')dx'dy' \le \pi\delta^2 [\varepsilon - \log r_0]$$

for any $\nu \ge \nu_0(\varepsilon,\delta)$. But $b_\nu \le 0$ on $\Omega$, therefore for any $\mu < \delta$ and any fixed $w$ with $|w - z| < \mu$ we have

$$\int_{|w-z'|<\delta+\mu} b_\nu(x',y')dx'dy' \le \int_{|z'-z|<\delta} b_\nu(x',y')dx'dy' .$$

Hence, for any $\delta,\varepsilon,\mu,w$ with $0 < \mu < \delta < \frac{1}{3}R$ and $|w - z| < \mu$ and $\varepsilon > 0$

$$\frac{1}{\pi(\delta + \mu)^2} \int_{|w-z'|<\delta+\mu} b_\nu(x',y')dx'dy' \le \frac{\delta^2}{(\delta + \mu)^2} [\varepsilon - \log r_0]$$

holds when $\nu \ge \nu_0(\varepsilon,\delta)$ .

Taking $\mu$ small enough, say $\mu = \mu(\varepsilon,\delta)$, we obtain for any $w$ such that $|w - z| < \mu$

$$\frac{1}{\pi(\delta + \mu)^2} \int_{|w-z'|<\delta+\mu} b_\nu(x',y')dx'dy' \le 2\varepsilon - \log r_0$$

when $\nu \ge \nu_0(\varepsilon,\delta)$.

Now $z \in \Omega$ is arbitrary and $\mu(\varepsilon,\delta)$ is independent of $z$. Hence by the compactness of $K$ there exists an index $\nu(\varepsilon,\delta,K)$ such that

$$\frac{1}{\pi(\delta + \mu)^2} \int_{|z-z'|<\delta+\mu} b_\nu(x',y')dx'dy' \le 2\varepsilon - \log r_0$$

when $\nu \ge \nu(\varepsilon,\delta,K)$ and $z \in K$.

By assumption the functions $a_\nu$, $\nu = 1,2,3,\ldots$, are analytic on $\Omega$ and this fact is used to obtain from the above integral estimate a pointwise estimate for the functions $b_\nu$. As is well-known, the logarithm of the modulus of an analytic function is subharmonic (see e.g. HÖRMANDER [2], p. 18). So

$$b_\nu(z) \le \frac{1}{\pi(\delta+\mu)^2} \int_{|z-z'|<\delta+\mu} b_\nu(x',y')dx'dy',$$

from which in combination with the above integral estimate it follows that

$$b_\nu(z) \le 2\varepsilon - \log r_0$$

when $\nu \ge \nu(\varepsilon,\delta,K)$ and $z \in K$. The last inequality can be rewritten into

$$\left[\frac{r_0}{e^{2\varepsilon}}\right]^\nu |a_\nu(z)| \le M .$$

Hence if we take $\varepsilon$ such that $\frac{r_0}{e^{2\varepsilon}} = r$ then

$$r^\nu |a_\nu(z)| \le M,$$

when $\nu \ge \nu(r,\delta,K)$ and $z \in K$, which completes the proof of the theorem.

Moreover, we need the following form of Schwarz' lemma.

Lemma 1. Let $f$ be analytic on the disc $\Omega = \{z \mid |z| \le R\}$ and let $|f| \le M$ on $\Omega$.

Then

$$|f(z) - f(w)| \le \frac{2M}{R - |z|} |z - w|$$

when $z,w \in \Omega$.

Proof. Apply the Schwarz' lemma in the usual form (see Rudin [3], p. 241) to the function $g$, defined on the disc $\{h \mid |h| < R - |z|\}$ by

$$g_z(h) = f(z) - f(z + h).$$

## §3. THE CASE $L = \Delta$ (2 DIMENSIONS)

In this section we consider a function u of two independent variables z and x, defined on an open subset $\Omega \subset \mathbb{C} \times \mathbb{R}^2$, such that u is analytic in z when x is given an arbitrary fixed value and such that u is harmonic when z is given an arbitrary fixed value. We shall give a complete proof that such a function u has to be regular on $\Omega$ in both variables together. More precisely, we shall prove the following theorem.

Theorem 2. Let u be a complex-valued function of two independent variables z and x, defined on an open subset $\Omega \subset \mathbb{C} \times \mathbb{R}^2$ such that u is $C^1$ in z when x is given an arbitrary fixed value, $C^2$ in x when z is given an arbitrary fixed value and satisfies

$$\frac{\partial u}{\partial \bar{z}} = 0 \quad , \quad \Delta_x u = 0$$

on $\Omega$. Then u is $C^\infty$ on $\Omega$ in both variables together.

Proof.

a) If suffices to prove that u is continuous. More precisely, if u is continuous on an open subset $\omega \subset \Omega$. Then u is $C^\infty$ on $\omega \subset \Omega$.
For u satisfies $\frac{\partial u}{\partial \bar{z}} = 0$ and $\Delta_x u = 0$ and u is continuous on $\omega$, so u is a continuous distributional solution of the elliptic equation with constant coefficients

$$\left(\frac{\partial^2}{\partial z \partial \bar{z}} + \Delta_x\right) u = 0$$

on $\Omega$. Then by standard arguments (see e.g. YOSIDA [4] Corollary, p. 178) it follows that u is $C^\infty$.

b) Let $(z,x) \in \Omega$, arbitrary but fixed. By a) it is sufficient to prove the continuity of u in a neighborhood of (z,x). We may suppose that

$$D_1 \times D_2 \equiv \{(z,x) \mid |z| \le 1, \ |x| \le 1\} \subset \Omega$$

and $(z,x) = (0,0)$. Let $E_M$ be the set of points $x \in D_2$ such that

$|u(z,x)| \leq M$ for all $z \in D_1$. Then $E_M$ is closed and $\bigcup_{M=1}^{\infty} E_M = D_2$. Then Baire's theorem (see e.g. YOSIDA [4], p. 11) implies that some $E_M$ contains an inner point $\xi$. Hence for $0 < \mu < \frac{1}{2}$ small enough we have

$$\{x \mid |x - \xi| \leq \mu\} \subset E_M ,$$

that is $|u(z,x)| \leq M$ when $|z| \leq 1$ and $|x - \xi| \leq \mu$.

So u is bounded on the polydisc $\omega' = \{(z,x) \mid |z| \leq 1, |x - \xi| \leq \mu\}$.
We shall show that the last conclusion implies that u is also continuous in $\omega'$ in both variables z and x together. This can be seen as follows. We take $(z,x) \in \omega'$ arbitrary but fixed. Then

$$|u(z,x) - u(z',x')| \leq |u(z,x) - u(z,x')| + |u(z,x') - u(z',x')|$$

and by Schwarz' lemma (lemma 1, §2)

$$|u(z,x') - u(z',x')| \leq \frac{2M}{1 - |z|} |z - z'| .$$

So

$$|u(z,x) - u(z',x')| \leq |u(z,x) - u(z,x')| + \frac{2M}{1 - |z|} |z - z'| .$$

But u is continuous in x for fixed z. From this the continuity of u on $\omega'$ follows at once.

c) The next step is to show that continuity of u on a polydisc

$$\{(z,x) \mid |z| \leq R, |x - \xi| \leq \mu\} \subset D_1 \times D_2$$

implies the continuity of u on the polydisc

$$\{(z,x) \mid |z| \leq R - \delta, |x - \xi| \leq \mu + \rho\}$$

with $\delta > 0$ arbitrary small and $0 \leq \rho < 1 - \mu$.

Then by repeated use of this property a finite number of times it follows that u is continuous in a polydisc, which is a neighborhood of the origin

(0,0) and so we are done.

It is no restriction to suppose that $\xi = 0$. We introduce polar coordinates $x_1 = r\cos\phi$, $x_2 = r\sin\phi$ $(0 < r < 1-\mu)$ and $\tilde{u}(z;r,\phi) = u(z,x)$. It follows that $\tilde{u}$ is $C^\infty$ in $\phi$ and periodic with period $2\pi$. Hence we can write $\tilde{u}$ into a Fourier series

$$\tilde{u}(z;r,\phi) = \tfrac{1}{2}A_0(z,r) + \sum_{\nu=1}^{\infty} A_\nu(z,r)\cos\nu\phi + B_\nu(z,r)\sin\nu\phi$$

convergent for $r < 1-\mu$, where

$$A_\nu(z,r) = \frac{1}{\pi}\int_{-\pi}^{\pi} \tilde{u}(z;r,\phi)\cos\nu\phi\, d\phi$$

$$B_\nu(z,r) = \frac{1}{\pi}\int_{-\pi}^{\pi} \tilde{u}(z;r,\phi)\sin\nu\phi\, d\phi$$

Now

$$\Delta_x = \frac{\partial^2}{\partial r^2} + \frac{1}{r}\frac{\partial}{\partial r} + \frac{1}{r^2}\frac{\partial^2}{\partial\phi^2}$$

and it follows that

$$0 = \frac{1}{\pi}\int_{-\pi}^{\pi}(\Delta_x\tilde{u})\cos\nu\phi\, d\phi = \frac{1}{\pi}\int_{-\pi}^{\pi}\left(\frac{\partial^2}{\partial r^2} + \frac{1}{r}\frac{\partial}{\partial r} + \frac{1}{r^2}\frac{\partial^2}{\partial\phi^2}\right)\tilde{u}\cos\nu\phi\, d\phi$$

$$= \left(\frac{\partial^2}{\partial r^2} + \frac{1}{r}\frac{\partial}{\partial r} - \frac{\nu^2}{r^2}\right)\frac{1}{\pi}\int_{-\pi}^{\pi}\tilde{u}\cos\nu\phi\, d\phi = \left(\frac{\partial^2}{\partial r^2} + \frac{1}{r}\frac{\partial}{\partial r} - \frac{\nu^2}{r^2}\right)A_\nu.$$

Hence

$$\left(\frac{\partial^2}{\partial r^2} + \frac{1}{r}\frac{\partial}{\partial r} - \frac{\nu^2}{r^2}\right)A_\nu = 0$$

and in the same way

$$\left(\frac{\partial^2}{\partial r^2} + \frac{1}{r}\frac{\partial}{\partial r} - \frac{\nu^2}{r^2}\right)B_\nu = 0\ .$$

This is an ordinary differential equation and its general solution is

$$A_\nu(z,r) = \begin{cases} a_\nu(z)r^\nu + b_\nu^*(z)r^{-\nu} & \text{when } \nu \neq 0 \\ a_0(z) + b_0^*(z)\log r \end{cases}$$

$$B_\nu(z,r) = \begin{cases} b_\nu(z)r^\nu + a_\nu^* r^{-\nu} & \text{when } \nu \neq 0 \\ b_0(z) + a_0^*(z)\log r\,. \end{cases}$$

Now u is continuous in the points (z;0), therefore $\tilde{u}$ and consequently $A_\nu$ and $B_\nu$ must remain bounded as $r \downarrow 0$. This implies that

$$a_\nu^* = b_\nu^* = 0,\ \nu = 0,1,2,\dots\,.$$

Hence

$$\tilde{u}(z;r,\phi) = \tfrac{1}{2}a_0(z) + \sum_{\nu=1}^{\infty} r^\nu[a_\nu(z)\cos\nu\phi + b_\nu(z)\sin\nu\phi]\,,$$

convergent for $r < 1 - \mu$, where

$$r^\nu a_\nu(z) = \frac{1}{\pi}\int_{-\pi}^{\pi}\tilde{u}(z;r,\phi)\cos\nu\phi\,d\phi$$

$$r^\nu b_\nu(z) = \frac{1}{\pi}\int_{-\pi}^{\pi}\tilde{u}(z;r,\phi)\sin\nu\phi\,d\phi\,.$$

Now by a) it follows that u is $C^\infty$ in both variables together on the disc $\{(z,x) \mid |z| < R,\ |x| < \mu\}$. So it is permitted to differentiate under the integral sign and then we get $\frac{\partial}{\partial\bar{z}}a_\nu = \frac{\partial}{\partial\bar{z}}b_\nu = 0$, i.e. $a_\nu$ and $b_\nu$ are analytic on the disc $\{z \mid |z| < R\}$.
From Bessel's inequality (see COURANT-HILBERT [1], p. 51)

$$\tfrac{1}{2}|a_0(z)|^2 + \sum_{\nu=1}^{\infty} r^{2\nu}(|a_\nu(z)|^2 + |b_\nu(z)|^2) \le \frac{1}{\pi}\int_{-\pi}^{\pi}|\tilde{u}(z;r,\phi)|^2 d\phi$$

where $r < 1 - \mu$ and $M = \max\{|u(z;x)| \mid |z| \le R,\ |x| \le \mu\}$ it follows that

$$\lim_{\nu\to\infty} a_\nu(z)(\mu+\rho)^\nu = 0$$

$$\lim_{\nu\to\infty} b_\nu(z)(\mu+\rho)^\nu = 0$$

$$|a_\nu(z)\mu^\nu| \le 2M, \quad \nu = 0,1,2,\dots$$

$$|b_\nu(z)\mu^\nu| \le 2M, \quad \nu = 0,1,2,\dots\,,$$

provided $|z| \leq R$, $0 < \rho < 1 - \mu$.

Now fix $\delta > 0$ arbitrary small. By application of Hartog's theorem (theorem 1, §2) with $\Omega = \{z \mid |z| \leq R - \frac{1}{2}\delta\}$ and $K = \{z \mid |z| \leq R - \delta\}$ it follows that there exists an index $\nu(\delta)$ such that

$$|a_\nu(z)(\mu + \rho)^\nu| \leq 2M \quad \text{when} \quad |z| \leq R - \delta .$$

This implies that the series with continuous terms

$$\sum_{\nu=1}^{\infty} r^\nu [a_\nu(z) \cos \nu\phi + b_\nu(z) \sin \nu\phi]$$

converges uniformly on the polydisc

$$\{(z,x) \mid |z| \leq R - \delta, \ |x| \leq \mu + \rho'\} ,$$

where $\rho'$ is any number $0 < \rho' < \rho$ and hence

$$u(z;x) = \tilde{u}(z;r,\phi) = \tfrac{1}{2}a_0(z) + \sum_{\nu=1}^{\infty} r^\nu [a_\nu(z) \cos \nu\phi + b_\nu(z) \sin \nu\phi]$$

is continuous on $\{(z,x) \mid |z| \leq R - \delta, \ |x| \leq \mu + \rho'\}$ ,

which completes the proof.

## §4. THE CASE $L = \Delta^m$ (2 DIMENSIONS)

We shall need the following result.

Lemma 2. Let $L = (\frac{\partial}{\partial z_2})^k (\frac{\partial}{\partial \bar{z}_2})^\ell$ $(k \geq 1, \ \ell \geq 1)$ and suppose that any function of the two independent variables $z_1, z_2$, defined in an open subset $\Omega = \Omega_1 \times \Omega_2 \subset \mathbb{C}^2$ which is separately $C^\infty$ and satisfies the equations

$$\frac{\partial u}{\partial \bar{z}_1} = 0, \qquad Lu = 0$$

in $\Omega$, is automatically $C^\infty$ in $\Omega$ in both variables $z_1$ and $z_2$ together. Then given an arbitrary analytic function f in $\Omega$ it follows that any separately $C^\infty$ solution in $\Omega$ of

$$\frac{\partial u}{\partial \bar{z}_1} = 0, \qquad Lu = f$$

is in fact $C^\infty$ in $\Omega$ in both variables $z_1$ and $z_2$ together.

Proof. L as a differential operator with constant coefficients has a fundamental solution. By the definition of the space $H^s(\mathbb{R}^2)$ (YOSIDA [4], p. 155) and due to the fact that the Fouriertransform $\hat{\delta}$ of the Dirac-measure is a constant (YOSIDA [4], p. 152) it follows that $\delta \in H^{-s}(\mathbb{R}^2)$ for any $s > 1$. In particular $E \in L^2_{loc}(\Omega) = H^0_{loc}(\Omega)$. Therefore E is locally integrable in $\Omega$. Let $\xi \in \Omega_2$ be arbitrary but fixed and $\phi$ a $C^\infty$ real-valued function with compact support in $\Omega_2$ and such that $\phi$ equals 1 in a neighborhood of $\xi$ and consider the convolution $u_0 = E * (\phi f)$. Then $u_0$ is continuous in $\Omega$ in both variables together and from Morera's theorem it follows that $u_0$ is analytic in $z_1$, so $u_0$ is a continuous solution of

$$\frac{\partial v}{\partial \bar{z}_1} = 0, \qquad Lv = \phi f$$

and therefore a continuous distributional solution of the elliptic equation

$$\left[\left(\frac{\partial}{\partial \bar{z}_1} \frac{\partial}{\partial z_1}\right)^m + \left(\frac{\partial}{\partial \bar{z}_2} \frac{\partial}{\partial z_2}\right)^m\right]u = g,$$

where m is the maximum of k and $\ell$ and g is the $C^\infty$-function given by

$$g = \left(\frac{\partial}{\partial \bar{z}_2}\right)^{m-k}\left(\frac{\partial}{\partial z_2}\right)^{m-\ell}\phi f \ .$$

By standard arguments (see YOSIDA [4], Corollary p. 178) from the theory of elliptic operators it follows that $u_0$ is $C^\infty$ in both variables $z_1$ and $z_2$ together on $\Omega$. Let u be an arbitrary separately $C^\infty$-function on $\Omega$ which satisfies

$$\frac{\partial u}{\partial \bar{z}_1} = 0, \quad Lu = f .$$

Then the difference $u - u_0$ satisfies

$$\frac{\partial}{\partial \bar{z}_1}(u - u_0) = 0, \quad L(u - u_0) = 0$$

in an open set $\Omega_1 \times \omega$, where $\omega$ is a neighborhood of $\xi$ in which $\phi$ equals 1. Hence by assumption $u - u_0$ is $C^\infty$ in both variables $z_1$ and $z_2$ together on $\Omega_1 \times \omega$, which implies knowing that $u_0$ is $C^\infty$ too on $\Omega_1 \times \omega$ that $u$ is $C^\infty$ on $\Omega_1 \times \omega$, which completes the proof.

Theorem 3. Let $u$ be a function of two independent complex variables $z_1$ and $z_2$, defined in an open set $\Omega \subset C^2$ such that $u$ is separately $C^\infty$ and satisfies

$$\frac{\partial u}{\partial \bar{z}_1} = 0, \quad \left(\frac{\partial}{\partial \bar{z}_2}\right)^k \left(\frac{\partial}{\partial z_2}\right)^\ell u = 0$$

$k \geq 1$, $\ell \geq 1$ in $\Omega$. Then $u$ is $C^\infty$ in $\Omega$ in both variables $z_1$ and $z_2$ together.

Proof. We shall use induction with respect to $k$ and $\ell$. The theorem is true when $k = 1$ and $\ell = 1$. Assume that the theorem has already been proved for $k$ and $\ell$. Then in the first place we shall prove that the theorem is also true for $k + 1$ and $\ell$. Let $L$ be as in lemma 2, then

$$\frac{\partial u}{\partial \bar{z}_1} = 0, \quad \frac{\partial}{\partial \bar{z}_2} Lu = 0 .$$

It is no restriction to suppose that

$$D_1 \times D_2 \equiv \{(z_1,z_2) \mid |z_1| \leq 1, |z_2| \leq 1\} \subset \Omega$$

and to prove that $u$ is $C^\infty$ in both variables $z_1$ and $z_2$ together at the point $(0,0)$. In the same way as in the proof of theorem 2 (section 3) it follows by Baire's theorem (YOSIDA [4], p. 11) that $u$ is bounded in a polydisc $\omega' = \{(z_1,z_2) \mid |z_1| \leq 1, |z_2 - \xi| \leq \mu\}$ and consequently (by Schwarz' lemma) continuous in $\omega'$. Then by standard arguments about elliptic operators it follows that u is $C^\infty$ in $\omega'' = \{(z_1,z_2) \mid |z_1| < 1, |z_2 - \xi| < \mu\}$ . So Lu is $C^\infty$ on $\omega''$ and therefore

$$\frac{\partial u}{\partial \bar{z}_1} = 0$$

implies

$$\frac{\partial}{\partial \bar{z}_1} Lu = L \frac{\partial}{\partial \bar{z}_1} u = 0$$

from which together with $\frac{\partial}{\partial \bar{z}_2} Lu = 0$ it follows that Lu is separately analytic in $\omega''$.
By Hartogs theorem on separately analytic functions (see HÖRMANDER [2], p. 28) we know that Lu is analytic in $\omega''$. Now Lu is analytic in $z_2$ in $D_2 = \{z_2 \mid |z_2| \leq 1\}$ for any fixed $z_1 \in D_1 = \{z_1 \mid |z_1| \leq 1\}$. So we may expand Lu in a Taylor series around $\xi$

$$Lu\,(z_1,z_2) = \sum_{\nu=1}^{\infty} a_\nu(z_1)(z_2 - \xi)^\nu, \ |z_2 - \xi| < 1 - |\xi|,$$

where the functions $a_\nu$ are analytic on $\{z_1 \mid |z_1| < 1\}$ because of the analyticity of Lu in $\omega''$. Application of Hartogs' theorem on sequences of analytic functions (theorem 1, section 2) shows that Lu is analytic in the largest open polydisc around $(0,\xi)$ which is contained in $D_1 \times D_2$. Therefore by lemma 2 and the induction hypothesis it follows that u is $C^\infty$ on this largest open polydisc. Then in a finite number of steps it is proved that u is $C^\infty$ in a polydisc which is a neighborhood of (0,0). Induction with respect to $\ell$ is proved now very easily. We only have to apply the preceding result to the function $\tilde{u}$, defined by $\tilde{u}(z_1,z_2) = u(z_1,\bar{z}_2)$, which completes the proof.

## §4. THE CASE L = Δ (n-DIMENSIONS)

In case L equals the Laplacian Δ in the n-dimensional space we may state a theorem analogue to theorem 2 of section 3. This theorem may be proved along the same lines, but for instance be need spherical coordinates and expansions into spherical functions instead of polar coordinates and Fourier series.

REFERENCES

1. R. COURANT and D. HILBERT, Methods of Mathematical Physics, vol. 1, Interscience Publishers, 1953.

2. L. HÖRMANDER, An Introduction to Complex Analysis in Several Variables, North-Holland Publishing Company, 1973.

3. W. RUDIN, Real an Complex Analysis, McGraw-Hill, 1966.

4. K. YOSIDA, Functional Analysis, Springer-Verlag,1971.

Department of Mathematics
Delft University of Technology
The Netherlands

*Differential Equations and Applications*
*W. Eckhaus and E.M. de Jager (eds.)*


# On integral inequalities associated with ordinary regular differential expressions

R. J. Amos and W. N. Everitt

1. In this paper we are concerned with the integral inequality

$$\int_a^b \{p|f'|^2 + q|f|^2\} \geq \mu \int_a^b w|f|^2 \qquad (f \in D) \tag{1.1}$$

where p, q and w are real-valued coefficients on the closed bounded interval [a,b], with p and w non-negative, and D is a linear manifold of complex-valued functions on [a,b] chosen so that all the three integrals in (1.1) are absolutely convergent.

We are interested in the so-called regular case of this inequality, i.e. when the coefficients 1/p, q and w are all integrable (Lebesgue) on [a,b]. The use of the term regular in this case is in accordance with a similar usage of this word in the theory of ordinary differential operators which plays a fundamental rôle in determining the parameters of the inequality (1.1).

In the case of smooth coefficients the calculus of variations affords an important method of studying inequalities of the form (1.1). Taking the case of real-valued functions f on [a,b], which is clearly equivalent to the complex-valued case, the inequality may be written in the isoperimetric form

$$\int_a^b \{pf'^2 + qf^2\} \geq \mu \qquad \int_a^b wf^2 = 1. \qquad (f \in D) \tag{1.2}$$

This type of problem is classical and details may be found in the standard texts concerned with the calculus of variations; see Akhiezer [2, section A-33], Courant and Hilbert [7, chapter VI, section 1], Fomin and Gelfand [9, section 41] and Weinstock [15, chapter 8]. In all these cases the coefficients p, q and w are required to be continuous, with p continuously differentiable, on [a,b] and, depending on the method employed, the elements f of D may need to be twice continuously differentiable.

Some relaxation of the conditions required for the discussion of the inequalities (1.1) and (1.2) is given by Bradley and Everitt [6, theorem 1]. However as pointed out in [6, section 2, (2.1) and section 3, proof of theorem 3], the inequality in the regular case given there is not obtained under minimal conditions on the coefficients due to technical reasons in the nature of the proof used in [6].

The difficulty in proving (1.1) lies in the fact that whilst the inequality is required on the maximal set D, the parameters of the inequality, i.e. the best-possible value of $\mu$ and the resulting cases of equality, are determined by a self-adjoint differential operator T with domain D(T); it will be seen below that D(T) is a strict subset of D and the inequality has to be first proved on, and then extended from D(T) to D in such a way as not to disturb these parameters. In [6] this was achieved by showing that, in a suitable norm, the domain D(T) is dense in D; however this requires certain additional conditions on the coefficients p, q and w beyond the minimal conditions we are interested in in this paper. The methods used in [6] were extended by Amos and Everitt in [4] but still under similar conditions on the coefficients.

One method to avoid this difficulty is to use the idea of compact embedding of one Hilbert space into a larger Hilbert space; see the book by Adams [1, sections 2.20 and 2.21]. The possibility of using this idea was suggested by Penning and Sauer in their report [13] who based their method on the results of Hildebrandt [10] on quadratic forms.

The method of compact embedding in both the regular and singular case was developed by Amos in his PhD thesis [3], and in the singular case by Amos and Everitt [5].

In this paper we return to the regular case and show that a complete answer can be given to the problem raised by the inequality (1.1) under minimal conditions on the coefficients. As is to be expected the result which emerges takes the same form as the inequality under the smooth, and now classical, conditions. The Proof is based on the ideas discussed above, an inequality given by Everitt [8] and the theory of quasi-differential operators developed by Naimark in [12].

Before we state the result we give some notations. The symbol '$(x \in K)$' is to be read as 'for all elements of the set K'. The real line is denoted by R and $[a,b]$ is a compact interval of R. The complex field is denoted by C. AC denotes absolute continuity, and L Lebesgue integration. $L_w^2(a,b)$, where w is a non-negative weight function, represents the collection of Lebesque integrable-square functions with respect to w on $[a,b]$; $(\cdot,\cdot)_w$ and $\|\cdot\|_w$ represent the norm and inner-product when this integration space is regarded as a Hilbert function space; in order to meet with notations used in [10] and [13] the space $L_w^2(a,b)$ is also represented by the symbol $H_0$ with $(\cdot,\cdot)_0$ and $\|\cdot\|_0$ as inner-product and norm respectively.

The conditions to be satisfied by the coefficients p, q and w are

$$\begin{aligned}
&\text{(i)}\ p,\ q,\ w : [a,b] \to R \\
&\text{(ii)}\ p(x) > 0 \quad (\text{almost all } x \in [a,b]) \text{ and } 1/p \in L(a,b) \qquad (1.3)\\
&\text{(iii)}\ q \in L(a,b) \\
&\text{(iv)}\ w(x) > 0 \quad (\text{almost all } x \in [a,b]) \text{ and } w \in L(a,b).
\end{aligned}$$

The linear manifold D of $L_w^2(a,b)$ is defined by

$$D = \{f : [a,b] \to C : f \in AC[a,b] \text{ and } p^{1/2}f' \in L^2(a,b)\}. \tag{1.4}$$

Note that $|q|^{1/2}f$ and $w^{1/2}f \in L^2(a,b)$ $(f \in D)$ in view of (1.4) and (iii) and (iv) of (1.3); thus all three integrals in (1.1) above are absolutely convergent when $f \in D$.

The self-adjoint differential operator $T : D(T) \subset L_w^2(a,b) \to L_w^2(a,b)$ is defined by

$$D(T) = \{f : [a,b] \to C : f \in AC[a,b],\ pf' \in AC[a,b],$$
$$w^{-1}(-(pf')' + qf) \in L_w^2(a,b), \text{ and } (pf')(a) = (pf')(b) = 0\} \tag{1.5}$$

where $' \equiv d/x$ and $w^{-1}$ is the reciprocal function $1/w$, and

$$Tf = w^{-1}(-(pf')' + qf) \qquad (f \in D(T)). \tag{1.6}$$

Note that quasi-derivatives are involved in both these definitions; see [12, sections 15, 16 and 17].

It is shown in [12, section 17] that $T$ is self-adjoint in $L_w^2(a,b)$, and in [12, section 19.2] that $T$ has a discrete spectrum $\{\lambda_n : n = 1,2,3\ldots\}$ (say); all the eigenvalues $\{\lambda_n\}$ are simple and we denote the corresponding eigenvectors (eigenfunctions) by $\{\psi_n : n = 1,2,3,\ldots\}$. The operator $T$ is bounded below in $L_w^2(a,b)$ (it is here that the condition $p > 0$ is essential), i.e. $-\infty < \lambda_n < \lambda_{n+1} < \infty$ $(n = 1,2,3,\ldots)$ and

$$(Tf,f)_w \geq \lambda_1(f,f)_w \qquad (f \in D(T)), \tag{1.7}$$

with equality if and only if $f$ belongs to the eigenspace of $T$ at $\lambda_1$; for these results see [12, section 19.4, theorem 5] or [8, theorem 2], and for the inequality (1.7) the book by Kato [11, section 10, page 278].

An integration by parts and use of the boundary conditions in (1.5) shows that

$$(Tf,f)_w = \int_a^b w[w^{-1}(-(pf')' + qf)]\overline{f}$$

$$= -pf'.\overline{f}\Big|_a^b + \int_a^b \{p|f'|^2 + q|f|^2\}$$

for all $f \in D(T)$. This result and (1.7) yield the inequality

$$\int_a^b \{p|f'|^2 + q|f|^2\} \geq \lambda_1 \int_a^b w|f|^2 \qquad (f \in D(T)) \tag{1.8}$$

with equality if and only if f is in the eigenspace of T at $\lambda_1$. Note that this result implies that

$$D(T) \subset D \tag{1.9}$$

and that strict inclusion is implied. However (1.9) also follows directly from the definitions (1.4) and (1.5) and the condition $1/p \in L(a,b)$ of (ii) of (1.3).

The problem now is to extend the inequality (1.8) from D(T) to D and to determine whether or not such an extension, which would yield the desired inequality (1.1), involves changing $\lambda_1$ to a new number $\mu$ and introducing new cases of equality. The answer here is in the negative, i.e. $\mu = \lambda_1$ and no new cases of equality are introduced. We state this result as

*Theorem* *Let the coefficients* p, q *and* w *satisfy the conditions* (1.3); *let the linear manifold* D *of* $L_w^2(a,b)$ *be defined by* (1.4); *let the differential operator* T *with domain* $D(T) \subset D \subset L_w^2(a,b)$ *be defined by* (1.5); *then*

$$\int_a^b \{p|f'|^2 + q|f|^2\} \geq \lambda_1 \int_a^b w|f|^2 \qquad (f \in D) \tag{1.10}$$

*where* $\lambda_1$ *is the first eigenvalue of* T; *there is equality in* (1.10) *if and only if* f *is in the eigenspace of* T *at* $\lambda_1$.

*Proof*. See the sections below.

*Remark.* It is essential to this result that the operator T be determined by the boundary conditions given in (1.5), *i.e.*

$$(pf')(a) = (pf')(b) = 0;$$

the reason for this appears in the proof given below; see the end of section 5 below. See also the remarks made in [6, section 6], [4, theorem 2, remark 2] and [5, page 7].

*Acknowledgements* R. J. Amos acknowledges his indebtedness to the Science Research Council for financial support to enable him to undertake post-graduate research work in the University of Dundee during the academic years 1974-77.

2. We first give a result due to Everitt [8, Theorem 1] which is used later in the proof of the Theorem given in section 1.

*Lemma* 1. *Let the coefficients* p, q *and* w *satisfy the conditions* (1.3); *let the linear manifold* D *of* $L_w^2(a,b)$ *be defined by* (1.4); then

(i) $f \in L_w^2(a,b) \quad (f \in D)$

(ii) $|q|^{1/2} f \in L^2(a,b) \quad (f \in D)$

(iii) <u>given any $\varepsilon > 0$ there exists a positive number</u> $A(\varepsilon)$

($\equiv A(\varepsilon, p, q, w)$) such that

$$\int_a^b |q||f|^2 \leq \varepsilon \int_a^b p|f'|^2 + A(\varepsilon) \int_a^b w|f|^2 \qquad (f \in D).$$

<u>Proof</u>. See Everitt [8]. We note that (i) and (ii) have been shown to hold already; see section 1 above.

3. We now let $H_0$ denote the Hilbert function space $L_w^2(a,b)$ and let $H_1$ denote the Hilbert function space defined by

$$H_1 \overset{\text{def}}{=\!=} \{D \text{ endowed with the inner product}$$

$$(f,g)_1 = \int_a^b \{pf'\overline{g}' + wf\overline{g}\} \qquad (f,g \in D)\}.$$

Here $(\cdot,\cdot)_1$ and $\|\cdot\|_1$ denote the inner product and norm in $H_1$ respectively. We note that $H_1$ is indeed a Hilbert function space; see the argument in [13, Theorem 1].

It follows from Lemma 1 that if $f \in H_1$ then $f \in L_w^2(a,b) = H_0$ and hence that

$$H_1 \subset H_0.$$

Also since

$$\|f\|_1 = \int_a^b \{p|f'|^2 + w|f|^2\} \qquad (f \in H_1)$$

and

$$\|f\|_0 = \int_a^b w|f|^2 \qquad (f \in H_0)$$

then

$$\|f\|_0 \leq \|f\|_1 \qquad (f \in H_1 \subset H_0).$$

Thus there is a natural embedding of $H_1$ in $H_0$; see [1, section 2.21].

The sesquilinear form $Q : H_1 \times H_1 \to C$ is defined by

$$Q(f,g) \overset{\text{def}}{=\!=} \int_a^b \{pf'\bar{g}' + qf\bar{g}\} \qquad (f,g \in H_1),$$

with corresponding quadratic form $\hat{Q} : H_1 \to C$ defined by

$$\hat{Q}(f) \overset{\text{def}}{=\!=} \int_a^b \{p|f'|^2 + q|f|^2\} \qquad (f \in H_1).$$

It follows from Lemma 1 that the sesquilinear form $Q$ is well defined on $H_1 \times H_1$ and hence that the corresponding quadratic form $\hat{Q}$ is also well defined on $H_1$.

<u>Lemma 2</u>. <u>Let the coefficients</u> $p$, $q$ <u>and</u> $w$ <u>satisfy the conditions</u> (1.3); <u>let the Hilbert spaces</u> $H_0$ <u>and</u> $H_1$ <u>and the sesquilinear form</u> $Q$ <u>with corresponding quadratic form</u> $\hat{Q}$ <u>be defined as above</u>; <u>then</u>

(i) $Q$ <u>is bounded on</u> $H_1$, <u>i.e. there is a positive number</u> $K(\equiv K(p, q, w)$ <u>only</u>) <u>such that</u>

$$|Q(f,g)| \leq K\|f\|_1\|g\|_1 \qquad (f,g \in H_1)$$

(ii) $\hat{Q}$ <u>is coercive on</u> $H_1$ embedded in $H_0$, <u>i.e</u>. <u>there is a positive number</u> $L(= L(p, q, w)$ <u>only</u>) <u>such that</u>

$$\hat{Q}(f) = Q(f,f) \geq \tfrac{1}{2}\|f\|_1^2 - L\|f\|_0^2 \qquad (f \in H_1).$$

<u>Proof</u>. We have

$$Q(f,g) = \int_a^b \{pf'\bar{g}' + qf\bar{g}\}$$

$$= \int_a^b \{pf'\bar{g}' + wf\bar{g} - wf\bar{g} + qf\bar{g}\} \qquad (f,g \in H_1),$$

and hence

$$|Q(f,g)| \le \left|\int_a^b \{pf'\bar{g}' + wf\bar{g}\}\right| + \left|\int_a^b wf\bar{g}\right| + \left|\int_a^b qf\bar{g}\right|$$

$$\le \left|\int_a^b \{pf'\bar{g}' + wf\bar{g}\}\right| + \left|\int_a^b wf\bar{g}\right| + \int_a^b |qf\bar{g}|$$

$$= |(f,g)_1| + |(f,g)_0| + \int_a^b |q||f\bar{g}| \qquad (f,g \in H_1).$$

Then using the Cauchy-Schwarz inequality and Lemma 1 with $\varepsilon = \frac{1}{2}$ we obtain

$$|Q(f,g)| \le \|f\|_1\|g\|_1 + \|f\|_0\|g\|_0 + \left(\int_a^b |q||f|^2 \int_a^b |q||g|^2\right)^{1/2}$$

$$\le 2\|f\|_1\|g\|_1 + \left\{\left(\frac{1}{2}\int_a^b p|f'|^2 + A(\tfrac{1}{2})\int_a^b w|f|^2\right)\left(\frac{1}{2}\int_a^b p|g'|^2 + A(\tfrac{1}{2})\int_a^b w|g|^2\right)\right\}^{1/2}$$

$$\le 2\|f\|_1\|g\|_1 + (\tfrac{1}{2} + A(\tfrac{1}{2}))\ \|f\|_1\|g\|_1$$

$$= (\tfrac{5}{2} + A(\tfrac{1}{2}))\ \|f\|_1\|g\|_1 \qquad (f,g \in H_1).$$

This proves (i) of Lemma 2.

To prove (ii) we have

$$\hat{Q}(f) = \int_a^b \{p|f'|^2 + q|f|^2\}$$

$$= \int_a^b \{p|f'|^2 + w|f|^2 - w|f|^2 + q|f|^2\} \qquad (f \in H_1),$$

and hence

$$\hat{Q}(f) \geq \|f\|_1^2 - \|f\|_0^2 - \int_a^b |q||f|^2 \qquad (f \in H_1).$$

Then from Lemma 1 with $\varepsilon = \frac{1}{2}$ we obtain

$$\hat{Q}(f) \geq \|f\|_1^2 - \|f\|_0^2 - \frac{1}{2}\int_a^b p|f'|^2 - A(\tfrac{1}{2})\int_a^b w|f|^2$$

$$\geq \frac{1}{2}\|f\|_1^2 - (1 + A(\tfrac{1}{2}))\|f\|_0^2 \qquad (f \in H_1).$$

This proves (ii) of Lemma 2.

4. We now have

Lemma 3. Let the coefficients p, q and w satisfy the conditions (1.3). Then the natural embedding of $H_1$ in $H_0$ is compact.

Remark. In the context of this paper the concept of compact embedding is best seen in the light of the remarks made in Hildebrandt [10, Section 2, Conditions (A) and (B)]; see also the book by Adams [1, Sections 2.20 and 2.21] and the report of Penning and Sauer [13, Page 1]. In particular we note that in the terms of [1, Section 1.23]:

(i) $H_1$ is embedded in $H_0$ since $H_1$ is a subset of $H_0$ and $\|f\|_0 \leq \|f\|_1$ $(f \in H_1)$;

(ii) with (i) satisfied it is sufficient, in order to obtain the compact embedding of $H_1$ in $H_0$, to show that any bounded subset of $H_1$ is precompact in $H_0$ and this we do in the proof of Lemma 3.

Proof. To prove Lemma 3 we sue the Riesz conditions for a subset K of $L^2(-\infty,\infty)$ to be precompact in $L^2(-\infty,\infty)$; for these conditions see [1, Sections 2.20 and 2.21], i.e. let K be a bounded subset of $L^2(-\infty,\infty)$, say (here $\|\cdot\|$ represents the usual norm in $L^2(-\infty,\infty)$)

$$\|F\| \leq B < \infty \qquad (F \in K) \tag{4.1}$$

<u>for some positive number</u> B; <u>then</u> K <u>is precompact in</u> $L^2(-\infty,\infty)$ <u>if and only if</u>

(α) <u>uniformly for all</u> $F \in K$

$$\lim_{X\to\infty} \int_X^\infty |F(x)|^2 dx = 0 \quad \underline{\text{and}} \lim_{X\to-\infty} \int_{-\infty}^X |F(x)|^2 dx = 0$$

(β) <u>uniformly for all</u> $F \in K$

$$\lim_{h\to 0} \int_{-\infty}^\infty |F(x+h) - F(x)|^2 dx = 0.$$

Define the subset K' of $H_1$ by, for some positive number B,

$$K' = \{f \in H_1 : \|f\|_1 \leq B\}.$$

Define the elements f of $H_1$ to satisfy the conditions

$$f(x) = 0 \qquad (x \in (-\infty,a) \cup (b,+\infty)). \tag{4.2}$$

Then define K as a subset of $L^2(-\infty,\infty)$ by

$$K = \{F = w^{1/2} f \quad \text{for all } f \in K'\}.$$

Condition (4.1) is satisfied for K since $\|f\|_0 \leq \|f\|_1 \quad (f \in H_1)$. If it can be shown that K is precompact in $L^2(-\infty,\infty)$ then a straightforward argument (recalling that w is positive almost everywhere on (a,b)) shows that K' is precompact in $H_0 = L^2_w(a,b)$. This then implies, see the remark given after the statement of Lemma 3, that $H_1$ is compactly embedded in $H_0$.

Since $-\infty < a < b < +\infty$ and condition (4.1) holds for all $f \in H_1$ it is clear that (α) is satisfied for all $F \in K$.

We now prove (β). Define the functions p and w on $(-\infty,a) \cup (b,\infty)$ by

$$p(x) = w(x) = 0 \qquad (x \in (-\infty,a) \cup (b,\infty)).$$

Now

$$f(x+h) - f(x) = \int_x^{x+h} f' \qquad (x,h \in R \text{ such that } x \text{ and } x+h \in [a,b];\ f \in K').$$

Taking moduli, squaring and using the Cauchy-Schwarz inequality this becomes

$$|f(x+h) - f(x)|^2 \le \left|\int_x^{x+h} p^{-1}\right| \left|\int_x^{x+h} p|f'|^2\right|$$

$$\le \left|\int_x^{x+h} p^{-1}\right| \|f\|_1^2 \qquad (x,h \in R \text{ such that } x \text{ and } x+h \in [a,b];\ f \in K'),$$

and hence

$$|f(x+h) - f(x)|^2 \le B'^2 \left|\int_x^{x+h} p^{-1}\right| \qquad (x,h \in R \text{ such that } x \text{ and } x+h \in [a,b];\ f \in K'). \tag{4.3}$$

We have

$$f(x) = f(t) + \int_t^x f' \qquad (x,t \in [a,b];\ f \in K').$$

Using the inequality $|\alpha + \beta|^2 \le 2|\alpha|^2 + 2|\beta|^2$ and the Cauchy-Schwarz inequality we see that

$$|f(x)|^2 \le 2|f(t)|^2 + 2\left|\int_t^x f'\right|^2$$

$$\le 2|f(t)|^2 + 2\left(\int_t^x |f'|\right)^2$$

$$\le |f(t)|^2 + 2 \int_a^b p^{-1} \int_a^b p|f'|^2 \qquad (x,t \in [a,b];\ f \in K').$$

Multiplying by w(t) and integrating with respect to t from a to b we obtain

$$|f(x)|^2 \int_a^b w \leq 2\int_a^b w|f|^2 + 2\int_a^b w \int_a^b p^{-1} \int_a^b p|f'|^2 \qquad (x \in [a,b];\ f \in K').$$

Thus

$$\begin{aligned} |f(x)|^2 &\leq 2\left(\int_a^b w\right)^{-1} \int_a^b w|f|^2 + 2\int_a^b p^{-1} \int_a^b p|f'|^2 \\ &\leq \left\{2\left(\int_a^b w\right)^{-1} + 2\int_a^b p^{-1}\right\}\left\{\int_a^b w|f|^2 + \int_a^b p|f'|^2\right\} \\ &\leq B'^2\left\{2\left(\int_a^b w\right)^{-1} + 2\int_a^b p^{-1}\right\} \qquad (x \in [a,b];\ f \in K'). \end{aligned} \tag{4.4}$$

From (4.4) we obtain

$$\int_x^b w|f|^2 \leq B'^2\left\{2\left(\int_a^b w\right)^{-1} + 2\int_a^b p^{-1}\right\}\int_x^b w \qquad (x \in [a,b];\ f \in K'). \tag{4.5}$$

Now the right hand side of (4.5) is independent of $f \in K'$ and tends to zero as $x \to b-$. Thus

$$\lim_{x\to b-} \int_x^b w|f|^2 = 0 \quad \text{uniformly for all } f \in K'. \tag{4.6}$$

Similarly it can be shown that

$$\lim_{x\to a+} \int_a^x w|f|^2 = 0 \quad \text{uniformly for all } f \in K'. \tag{4.7}$$

Then for all $h \in R$ and all $f \in K'$

$$\begin{aligned} &\int_{-\infty}^{\infty} |w^{1/2}(x + h)f(x + h) - w^{1/2}(x)f(x)|^2 dx \\ &\leq 2\int_{-\infty}^{\infty} |w^{1/2}(x + h) - w^{1/2}(x)|^2 |f(x + h)|^2 dx + 2\int_{-\infty}^{\infty} w(x)|f(x + h) - f(x)|^2 dx. \end{aligned} \tag{4.8}$$

We note that all the integrals in (4.8) converge since $f \in AC[a,b]$, $w \in L(a,b)$ and both $f$ and $w$ are null outside $[a,b]$. From (4.4) we see that for all $h \in R$ and $f \in S'$

$$\int_{-\infty}^{\infty} |w^{1/2}(x + h) - w^{1/2}(x)|^2 |f(x + h)|^2 dx$$

$$\leq 2B'^2 \left\{ \left( \int_a^b w \right)^{-1} + \int_a^b p^{-1} \right\} \int_{-\infty}^{\infty} |w^{1/2}(x + h) - w^{1/2}(x)|^2 dx.$$

Now $w^{1/2} \in L^2(-\infty,\infty)$ and so from a general theorem for the Lebesgue integral, see Titchmarsh [14, Page 397, Example 19], we have

$$\lim_{h\to 0} \int_{-\infty}^{\infty} |w^{1/2}(x + h) - w^{1/2}(x)|^2 dx = 0.$$

Thus

$$\lim_{h\to 0} \int_{-\infty}^{\infty} |w^{1/2}(x + h) - w^{1/2}(x)|^2 |f(x + h)|^2 dx = 0 \text{ uniformly for all } f \in K'. \tag{4.9}$$

From (4.3) we see that for all $h \in R$ and all $f \in K'$

$$\int_{-\infty}^{\infty} w(x)|f(x + h) - f(x)|^2 dx$$

$$\leq 2B'^2 \int_a^b w(x) \left| \int_x^{x+h} p^{-1} \right| dx + \int_{b-|h|}^b w|f|^2 + \int_a^{a+|h|} w|f|^2. \tag{4.10}$$

Now

$$\lim_{h\to 0} \left| \int_x^{x+h} p^{-1} \right| = 0 \qquad (x \in [a,b]),$$

$$\left| \int_x^{x+h} p^{-1} \right| \leq \int_x^b p^{\prime 1} \qquad (x \in [a,b];\ h \geq 0),$$

and

$$\left| \int_x^{x+h} p^{-1} \right| \leq \int_a^x p^{-1} \qquad (x \in [a,b];\ h \leq 0),$$

so that by dominated convergence for the Lebesgue integral it follows that

$$\lim_{h\to 0+} \int_a^b w(x)\left|\int_x^{x+h} p^{-1}\right| dx = \lim_{h\to 0-} \int_a^b w(x)\left|\int_x^{x+h} p^{-1}\right| dx = 0. \tag{4.11}$$

Thus from (4.6), (4.7), (4.10) and (4.11) it follows that

$$\lim_{h\to 0} \int_{-\infty}^{\infty} w(x)|f(x+h) - f(x)|^2 dx = 0 \text{ uniformly for all } f \in K'. \tag{4.12}$$

Then from (4.8), (4.9) and (4.12) it follows that

$$\lim_{h\to 0} \int_{-\infty}^{\infty} |w^{1/2}(x+h)f(x+h) - w^{1/2}(x)f(x)|^2 dx = 0 \text{ uniformly for all } f \in K'.$$

It now follows that condition ($\beta$) is satisfied. This completes the proof of Lemma 3.

Corollary to Lemma 3. *Let the coefficients* $p$, $q$ *and* $w$ *satisfy the conditions* (1.3); *let* $\mu$ *be defined by*

$$\mu \overset{\text{def}}{=} \inf \{\hat{Q}(f) : f \in H_1 \text{ with } \|f\|_0 = 1\} \tag{4.13}$$

*then*

(a) $\mu > -\infty$

(b) *there is a vector* $g \in H_1$ *such that*

$\|g\|_0 = 1 \qquad Q(g,g) = \mu$

*and* $$Q(f,g) = \mu(f,g)_0 \qquad (f \in H_1). \tag{4.14}$$

*Proof*. (a) The lower bound $\mu$ of $\{\hat{Q}(f) : f \in H_1 \text{ with } \|f\|_0 = 1\}$ must be finite since $\hat{Q}$ is coercive on $H_1$ embedded in $H_0$ from (ii) of Lemma 2.

(b) The existance of a vector $g \in H_1$ satisfying all the required properties follows from a result in the general theory of quadratic forms in Hilbert space as given in the paper of Hildebrandt, see [10, Page 417]. The application of the results in [10] requires essentially the compact embedding of $H_1$ in $H_0$.

5. We now prove the Theorem given in section 1 which identifies the parameters $\mu$ and $g$ of the Corollary to Lemma 3 with $\lambda_1$ and $\psi_1$ (say), respectively the first eigenvalue and a corresponding normalised $(\|\psi_1\|_0 = 1)$ eigenfunction of the operator T.

We have, from (1.8) and the definition of $\hat{Q}$, that

$$\hat{Q}(f) = \int_a^b \{p|f'|^2 + q|f|^2\} \geq \lambda_1 \int_a^b w|f|^2 = \lambda_1(f,f)_0 \qquad (f \in D(T)).$$

Now since $D(T) \subset H_1$ it follows from the definition (4.13) of $\mu$ that

$$\mu \leq \lambda_1. \tag{5.1}$$

To prove that $\mu = \lambda_1$ we use the result (4.14) which, in view of the definition of Q, may be written in the form

$$\int_a^b \{pf'\bar{g}' + qf\bar{g}\} = \mu \int_a^b wf\bar{g} \qquad (f \in H_1). \tag{5.2}$$

Since $D(T) \subset H_1$ it follows that (5.2) is valid for all $f \in D(T)$. Then upon an integration by parts, noting that $(pf')(a) = (pf')(b) = 0 \quad (f \in D(T))$, we obtain

$$\int_a^b w\, Tf\, \bar{g} = \mu \int_a^b w\, f\, \bar{g} \qquad (f \in D(T)), \tag{5.3}$$

i.e.

$$(Tf,g)_0 = \mu(f,g)_0 \qquad (f \in D(T)),$$

and hence

$$((T-\mu)f,g)_0 = 0 \qquad (f \in D(T)).$$

Thus the vector g is in the orthogonal complement in $L_w^2(a,b) = H_0$ of the range of the operator $T - \mu$. Since T is self-adjoint and $\|g\|_0 = 1$ it follows from standard properties of self-adjoint operators in Hilbert space that $\mu$ is an eigenvalue of T and that g is in the eigenspace of T at the eigenvalue $\mu$. However $\mu \le \lambda_1$ from (5.1) and so since $\lambda_1$ is the smallest eigenvalue of T it follows that

$$\mu = \lambda_1,$$

and that g is in the eigenspace of T at $\lambda_1$.

Note that in passing from (5.2) to (5.3) we have used the boundary conditions in the definition of the operator T in (1.5). These particular boundary conditions are essential since, as we see from above, g is a non-null solution of the differential equation (since it is an eigenvector of T with $\|g\|_w = 1$)

$$-(py')' + qy = \lambda wy \text{ on } [a,b]$$

with $(pg')(a) = (pg')(b) = 0$ i.e. $g(a) \neq 0$ and $g(b) \neq 0$. The operator T defined by (1.5) and (1.6) is unique in determining the correct parameter $\mu$ and the cases of equality in (1.1).

The proof of the Theorem given in section 1 now follows.

## References


1. R. A. Adams, Sobolev spaces (Academic Press, New York, 1975).

2. N. I. Akhiezer, The calculus of variations (Blaisdell, New York, 1962).

3. R. J. Amos, On some problems concerned with integral inequalities associated with symmetric ordinary differential expressions, Ph.D. thesis, University of Dundee, 1977.

4. R. J. Amos and W. N. Everitt, On a quadratic integral inequality, Proc. Roy. Soc. Edin. (to appear).

5. R. J. Amos and W. N. Everitt, On integral inequalities and compact embeddings associated with ordinary differential expressions (submitted for publication).

6. J. S. Bradley and W. N. Everitt, Inequalities associated with regular and singular problems in the calculus of variations, Trans. Amer. Math. Soc. 182 (1973) 303-321

7. R. Courant and D. Hilbert, Methods of mathematical physics (Interscience, New York, 1953).

8. W. N. Everitt, An integral inequality with an application to ordinary differential operators, Proc. Roy. Soc. Edin. (to appear).

9. S. V. Fomin and I. M. Gelfand, Calculus of variations (Prentice-Hall, London, 1963).

10. S. Hildebrandt, Rand-und Eigenwertaufgaben bei stark elliptischen Systemen linearer Differentialgleichungen, Math. Ann. 148 (1962) 411-429.

11. T. Kato, Perturbation theory for linear operators (Springer-Verlag, Berlin, 1966).

12. M. A. Naimark, Linear differential operators : Volume II (Ungar, New York, 1968).

13. F. Penning and N. Sauer, Note on the minimization of $\int_0^\infty [p(x)|f'(x)|^2 + q(x)|f(x)|^2]$, University of Pretoria, Department of Applied Mathematics, Research Report UP TW 2, 1976.

14. E. C. Titchmarsh, Theory of functions (University Press, Oxford, 1939).

15. R. Weinstock, Calculus of variations, with applications to physics and engineering (McGraw-Hill, New York, 1952).

Department of Mathematics
University of Dundee
DUNDEE
Scotland UK.

*Differential Equations and Applications*
*W. Eckhaus and E.M. de Jager (eds.)*
*©North-Holland Publishing Company (1978)*

# CAN WE FIND OUT THE TOPOLOGICAL SHAPE OF A PLANET FROM ITS ATLAS ?

Gaetano Fichera
University of Rome

This paper is dedicated to the memory of Jan Van der Corput.

Everybody knows that a compact orientable surface is topologically determined by its genus $g$. The problem which I shall consider in this paper is the following:

Suppose that the compact and orientable surface is given through one of its atlas, i.e. suppose that a (finite) set of maps covering the surface is given with the relevant "connecting homeomorphism". Can we compute the genus $g$ of the surface?

We shall assume that the differential structure introduced in the manifold by the given atlas is $\mathcal{C}^\infty$. The problem is a particular case of the following one, which, actually, is the one we shall consider in this paper.

Given an atlas (finite set of maps and connecting homeomorphism ) of a $\mathcal{C}^\infty$ differentiable, orientable and compact manifold $V^r$ of dimension $r \geq 1$, compute the Betti numbers of $V^r$.

In dealing with this rather unusual problem in Analysis, in order to avoid any misunderstanding, it seems to me necessary to say that the word "compute" must be understood in the sense of Numerical Analysis, i.e. to give a mathematical procedure which, no matter how analitically sophisticated, is such that, using only the "data" of the problem (i.e. the functions which give the connecting homeomorphism) it can be programmed on an automatic computer. On the other hand we shall not assume any "convenience hypothesis" like the one which consists in supposing that the maps of the atlas constitute a <u>simple covering</u> of $V^r$. In fact in this case the homology of $V^r$ is the same as the homology of the nerve of the covering. This is a classical result due to Leray. The circumstance that such kind of coverings exist [1] [1] is of no help for computational purposes. In fact we have to consider that not only simple coverings exist, but also triangulations ([2],pp.125-135) of the variety exist : to use a triangulation would make the problem trivial. However the mere <u>existence</u> of some mathematical object is, generally, something very different from the actual computations connected with this object. This point of view, although commonly accepted by people working in Analysis, in particular in Partial Differential Equations, could not be so familiar to scientists, even outstanding, working in other fields.

Let $M_1, M_2, \dots, M_q$ be (open) intervals of the cartesian space $X^r$. Let $M_{i1}, M_{i2}, \dots, M_{iq}$ be open sets (some eventually empty) contained in $M_i$ and such that $M_{ii} = M_i$. Denote by $\tau_{ik}$ a $\mathcal{C}^\infty$-homeomorphism of the closure $\bar{M}_{ik}$ of $M_{ik}$ into $\bar{M}_{ki}$. Suppose that

1) $\tau_{ii}$ = identity;

2) $\tau_{ik} = \tau_{ki}^{-1}$ ;

3) $x \in M_{ik}$, $\tau_{ik}(x_i) \in M_{hk} \Longrightarrow \tau_{hk}[\tau_{ih}(x_i)] = \tau_{ik}(x_i)$;

4) $\tau_{ik}(\partial M_{ik} \cap M_i) \subset \partial M_k$;

5) for every $x \in \partial M_i$, there exists $h$ such that

$$x_i \in \bar{M}_{ih}, \quad \tau_{ih}(x_i) \in M_h ;$$

6) if $x_i \equiv (x_i^1, \dots, x_i^r)$ is the point of $M_i$ and if we consider the jacobian matrix $\frac{\partial \tau_{ik}}{\partial x_i}$ we have

---

(1) Numbers in brackets refer to the bibliography at the end of this paper.

$$\det \frac{\partial \tau_{ik}}{\partial x_i} > 0.$$

The collection $[M_i ; \tau_{ik}]$ which is the "datum" of our problem is an atlas.

There exists a $\mathcal{C}^\infty$ differentiable compact oriented manifold $V^r$ (determined up to a $\mathcal{C}^\infty$ homeomorphism) which has $[M_i, \tau_{ik}]$ as its atlas (see [3] p.545).

If we consider the equivalence relation: $x_i \in M_i \simeq x_k \in M_k$ whenever $x_i = \tau_{ki}(x_k)$ [i.e. $x_k = \tau_{ik}(x_i)$], the point $x$ of $V^r$ is the equivalence class determined by $x_i$ ; $x_i$ is the <u>image</u> of $x$ in the map $M_i$. Let $\varphi_k(x^1, \ldots, x^r)$ be a real valued function belonging to $\mathcal{C}^\infty(X^r)$ and such that

$$\varphi_k(x^1, \ldots, x^r) \begin{cases} > 0 & \text{in } M_k, \\ = 0 & \text{in } X^r - M_k \end{cases} \qquad (k = 1, \ldots, q).$$

Define on $V^r$ the function

$$\tilde{\varphi}_k(x) \begin{cases} = \varphi_k(x_k) & \text{if } x_k \text{ is the image of } x \text{ in } M_k \\ = 0 & \text{elsewhere.} \end{cases}$$

We have $\tilde{\varphi}_k(x) \in \mathcal{C}^\infty(V^r)$. Set

$$\varphi(x) = \sum_{k=1}^{q} \tilde{\varphi}_k(x).$$

Since $\varphi(x) > 0$ on $V^r$, we can consider

$$\psi_k(x) = \frac{\tilde{\varphi}_k(x)}{\varphi(x)}$$

and we have

$$\sum_{k=1}^{q} \psi_k(x) = 1.$$

Thus we have a special partition of unity on $V^r$ such that <u>supp</u> $\psi_k(x)$ is contained in (the domain of $V^r$ having like image in the atlas) $M_k$.

Set

$$a_{ij}^{(h)}(x) = \psi_h(x)\,\delta_{ij}$$

and consider $a_{ij}^{(h)}(x)$ like the components of a covariant symmetric tensor. Set

$$a_{ij}(x) = \sum_{h=1}^{q} a_{ij}^{(h)}(x)$$

and

$$ds^2 = a_{ij}(x)\,dx^i dx^j.$$

Since the $a_{ij}(x)$ are the components of a covariant symmetric positive tensor, we have introduced in $V^r$ a Riemannian metrics.

Let $v$ be a $\mathcal{C}^\infty$ real $k$-form (i.e. differential form of degree $k$) which in a local coordinate system is represented by

$$v = \frac{1}{k!}\, v_{s_1 \dots s_k}\, dx^{s_1} \dots dx^{s_k}.$$

It is well known that <u>the differential</u> $dv$ of $v$ is the $(k+1)$-form which is represented by

$$dv = \frac{1}{k!}\, dv_{s_1 \dots s_k}\, dx^{s_1} \dots dx^{s_k}$$

$$= \frac{1}{(k+1)!} \left( \sum_{h=1}^{k+1} (-1)^{h-1} \frac{\partial v_{j_1 \dots j_{h-1} j_{h+1} \dots j_{k+1}}}{\partial x^{j_h}}\, dx^{j_1} \dots dx^{j_{k+1}} \right).$$

Moreover $ddv = 0$.

The <u>adjoint form</u> of the $k$-form $v$ is the $(r-k)$-form

$$* v = \frac{1}{(r-k)!}\, * v_{s_{k+1} \dots s_r}\, dx^{s_{k+1}} \dots dx^{s_r},$$

where

$$* v_{s_{k+1} \dots s_r} = \frac{1}{k!}\, \delta^{1 \dots r}_{s_1 \dots s_r} \left[\det(a_{ij})\right]^{\frac{1}{2}} a^{s_1 i_1} \dots a^{s_k i_k}\, v_{i_1 \dots i_k}.$$

$\{a^{hi}\}$ is the symmetric contravariant positive tensor associated to $a_{ij}$, i.e. $a^{hi} a_{ij} = \delta^h_j$.

The <u>co-differentiation</u> operator $\delta$ is defined by

$$\delta = (-1)^{r(k+1)+1} * d *$$

and maps the $k$-form $v$ into its <u>co-differential</u> $\delta v$ which is a $(k-1)$-form. We have $\delta\delta v = 0$.

The Laplace-Beltrami differential operator for $k$-forms is the following:

$$\Delta = d\delta + \delta d \quad (2)$$

(2) Actually one gets the classical Laplace-Beltrami operator for $0$-forms (i.e. scalar functions) replacing $\Delta$ by $-\Delta$ .

Let $I$ be the identity operator (for $k$-forms) and set $G = (\Delta + I)^{-1}$. If we consider the space $\mathcal{L}^2_k(V^r)$, i.e. the space of real $k$-forms which have locally $\mathcal{L}^2$ coefficients, we may introduce a Hilbert structure in this space by means of the following scalar product:

$$(u,v) = \int u \wedge * v,$$

where $\int$ stands for integration extended over the oriented manifold $V^r$. We can perform this integration in terms of our "data". In fact

$$\int u \wedge * v = \sum_{h=1}^{q} \int_{M_h} \psi_h \frac{1}{k!(r-k)!} \delta^{s_1 \dots s_r}_{1 \dots r} u_{s_1 \dots s_k} * v_{s_{k+1} \dots s_r} dx^1 \dots dx^r.$$

The operator $G$ is a positive compact operator (briefly PCO) in the Hilbert space $\mathcal{L}^2_k(V^r)$ (see [4], p.154). Because of the Hodge theorem ([5] p.159, [6] p.225) we have that the Betti number of the dimension $k$ : $b_k$ equals the geometric multiplicity of the largest "eigenvalue" $\mu = 1$ of the PCO $G$. It is understood that if $b_k = 0$ the "eigenvalue" $\mu = 1$ has geometric multiplicity zero, in other words $\mu = 1$ is not an eigenvalue for $G$.

As a consequence we may assert that the problem of the computation of $b_k$ is a particular case of the more general problem concerning the computation of the geometric multiplicity $p$ of the largest eigenvalue of a PCO.(3)

For the solution of this problem we shall use the Hilbert space group theoretic approach which has been used in eigenvalue theory in the papers [7], [8]. Following this approach we are led to consider a sequence of roto-homothetic invariants

(3) Actually our concern is a little more general. In fact we are interested in the following problem: let $\mu$ be a positive number not less than the maximum eigenvalue of a given PCO. Compute the geometric multiplicity $p$ of $\mu$, where $p = 0$ if $\mu$ is not an eigenvalue for $G$.

providing the computation of $p$.

Let $\mu_1 \geq \mu_2 \geq \cdots \geq \mu_k \geq \cdots$ be the sequence of the eigenvalues of the PCO $G$ of the Hilbert space $S$ each repeated according to its (geometric) multiplicity. Suppose that for some integer $n>0$ $G^n$ has a finite Hilbert-Schmidt trace. Set for $s=1,2,\ldots$

$$\mathfrak{I}_s^n(G) = \frac{(-1)^s}{2\pi i} \int_{+C} \frac{\prod_{k=1}^{\infty}(1-\lambda\mu_k^n)}{\lambda^{s+1}}\, d\lambda ,$$

where $C$ is any rectifiable contour of the complex plane enclosing the origin.

Set

$$\psi_n^{(s)}(G) = \frac{1}{\mu_1^{ns}} \mathfrak{I}_s^n(G) .$$

$\psi_n^{(s)}(G)$ is a roto-homothetic invariant (i.e.an orthogonal invariant of degree zero) such that

i) $\qquad \psi_n^{(s)}(G) > \psi_{n+1}^{(s)}(G)$ ;

ii) (1) $\qquad \lim_{n\to\infty} \psi_n^{(s)}(G) \begin{cases} =0 & \text{if } p<s , \\ =\binom{p}{s} & \text{if } p \geq s . \end{cases}$

(see [7] ,p.259).

To the actual computation of the orthogonal invariant $\mathfrak{I}_s^n(G)$ provides the following representation theorem (see [6] p.333)

(2) $$\mathfrak{I}_s^n(G) = \frac{1}{s!} \sum_{h_1,\ldots,h_s} \det\{(G^n u_{h_i}, u_{h_j})\} \qquad (i,j=1,\ldots,s)$$

where $u_1,\ldots,u_k,\ldots$ is an arbitrary complete orthonormal system in the space $S$.

Returning to the operator $G=(\Delta+I)^{-1}$, we know, from the theory of elliptic operators, that, for $n>\frac{1}{2}r$ , $G^n$ has a finite Hilbert-Schmidt trace, hence (2) applies. For simplicity we assume $n=2m$ , $s=1$ . From (1) we deduce

$$b_k = \lim_{m\to\infty} \mathfrak{I}_1^{2m}(G).$$

Consider the following $k$-form on $V^r$

$$\omega^{(\ell;i_1,\ldots,i_r;s_1,\ldots,s_k)} \begin{cases} = \psi_h(x^1,\ldots,x^r)(x^1)^{i_1}\ldots(x^r)^{i_r} dx^{s_1}\ldots dx^{s_k} \\ \qquad \text{if } x \in \text{ image of } M_\ell \text{ on } V^r ; \\ = 0 \quad \text{elsewhere} \end{cases}$$

$$(\ell = 1,\dots,q\ ;\ i_1,\dots,i_r = 0,1,2,\dots\ ;\ s_1,\dots,s_k = 1,\dots,r)$$

Fixed the integer $\nu > 0$, let us orthonormalize by the Gram-Schmidt procedure the sequence

$$\{(\Delta + I)^{\nu}\,\omega^{(\ell;\,i_1,\dots,i_r\,;\,s_1,\dots,s_k)}\}$$

$$(\ell = 1,\dots,q\ ;\ i_1,\dots,i_r = 0,1,2,\dots\ ;\ s_1,\dots,s_k = 1,\dots,r).$$

Denote by $\{(\Delta + I)^{\nu} v_h^{\nu}\}$ the sequence which has been obtained by the orthonormalization process. From (2) we deduce

$$\mathcal{J}_1^{2m}(G) = \sum_{h=1}^{\infty} \|(\Delta + I)^{\lambda} v_h^{m+\lambda}\|^2,$$

where $\lambda$ is any arbitrarily chosen non-negative integer. From (1), assuming $\lambda = 0$, we deduce the unexpectedly simple limit relation

$$\boxed{b_k = \lim_{m \to \infty} \sum_{h=1}^{\infty} \|v_h^m\|^2}$$

and moreover

$$\sum_{h=1}^{\infty} \|v_h^m\|^2 > \sum_{h=1}^{\infty} \|v_h^{m+1}\|^2 > b_k.$$

An analogous result has been obtained by M.P.Colautti [3] who proves a less elegant formula, which, however, is more suitable for numerical computations.

## REFERENCES


[1] A.WEYL, Sur les théorèmes de de Rham, Comm.Mathem.Helvetici, 26,1952,pp.119-145.

[2] H.WHITNEY, Geometric Integration Theory, Princeton Univ. Press, 1957.

[3] M.P.COLAUTTI, Sul calcolo dei numeri di Betti di una varietà differenziabile, nota per mezzo di un suo atlante, Rend.di Matem.22,1963,pp.543-556.

[4] G.FICHERA, Teoria assiomatica delle forme armoniche, Rend. di Matem.,20,1961,pp.147-171.

[5] G.DE RHAM, Variétés differéntiables, Hermann & C.ie Ed. Paris, 1955.

[6] G.FICHERA, Spazi lineari di k-misure e di forme differenziali, Proceed.of the Symp.on Linear Spaces (Jerusalem,1960) Jerusalem Acad.Press,Pergamon Press,London,1961,pp.175-226.

[7] G.FICHERA, Approximation and Estimates for Eigenvalues, Proc. of the Symp.on the Numerical Solution of PDE (University of Maryland,1965), Acad.Press, New York-London, 1966,pp.317-352.

[8] G.FICHERA, Invarianza rispetto al gruppo unitario e calcolo degli autovalori, Ist.Naz.di Alta Matem.Symposia Mathematica X, Academic Press New York,1972,pp.255-264.